Plant Types 1

Algae, fungi and lichens

Ruth N. Miller

HUTCHINSON

London Melbourne Sydney Auckland Johannesburg

Hutchinson & Co. (Publishers) Ltd

An imprint of the Hutchinson Publishing Group

17-21 Conway Street, London W1P 6JD

Hutchinson Group (Australia) Pty Ltd
30-32 Cremorne Street, Richmond South, Victoria 3121
PO Box 151, Broadway, New South Wales 2007

Hutchinson Group (NZ) Ltd
32-34 View Road, PO Box 40-086, Glenfield, Auckland 10

Hutchinson Group (SA) (Pty) Ltd
PO Box 337, Bergvlei 2012, South Africa

First published 1982
© Ruth N. Miller 1982

Set in 9/10 Univers by Illustrated Arts, Sutton, Surrey.

Printed in Great Britain by The Anchor Press Ltd
and bound by Wm Brendon & Son Ltd
both of Tiptree, Essex

British Library Cataloguing in Publication Data

Miller, Ruth N.
  Plant types.
  1: Algae, fungi and lichens
  1. Plants
  I. Title
  581   QK49

ISBN 0 09 144481 0

Contents

Contents

Contents

Algae

This is a large group of plants, showing wide variety in the organization of the plant body, reproductive processes and life cycles. The algae were formerly classified, together with the fungi and the lichens, in the division Thallophyta on the basis of the following characteristics:

plant body a simple, undifferentiated structure called a thallus

little internal cellular differentiation

usually unicellular sex organs.

More recently it has been suggested that the division Thallophyta be split and the algae classified into ten separate divisions. Nowadays, therefore, taxonomists do not consider the term 'alga' to have significance in a systematic classification, although it is still convenient to use the word for the seaweeds and their relatives generally. It is important to realize that vegetative organization and reproductive processes are of little use in the primary classification of the algae, because of the wide variety already mentioned. Five main criteria have been derived from electron microscope, biochemical and physiological studies. These are:

(a) photosynthetic pigments

(b) food storage products

(c) cell wall components

(d) flagella

(e) fine cell structure.

All the algae possess chlorophyll *a*, together with a variety of other chlorophylls, carotenoids and biloproteins, which can give distinctive colour to a particular group; e.g. fucoxanthin, a brown pigment, gives the characteristic colour to the Phaeophyta (brown algae).

Flagella are not found in the Cyanophyta or the Rhodophyta, but are useful characters for classification in all the other divisions, because their position, number and form vary from group to group. Algal flagella, when seen in section in electron microscope preparations, show the $9+2$ internal fibril pattern common to the cilia and flagella of other organisms, but their external appearance varies in that they are either smooth or whiplike (acronematic), or they have rows of very fine hairs arranged longitudinally giving a 'tinsel' effect (pantonematic).

Although the other diagnostic criteria are not as easily seen as some of the pigments or the type of flagella, the variations will be mentioned for the seven algal divisions to be described here.

The divisions are:

Division Cyanophyta – Blue-green algae

Division Chlorophyta – Green algae

Division Euglenophyta – Euglenoids

Division Bacillariophyta – Diatoms

Division Xanthophyta – Yellow-green algae

Division Phaeophyta – Brown algae

Division Rhodophyta – Red algae.

Division Cyanophyta
Blue-green algae

Marine (littoral (shore) zone and plankton), freshwater and terrestrial.
Some species live in hot springs, tolerating temperatures as high as 85°C. In the plankton, Cyanophyta are most evident during the warmer months of the year. One or two species may reproduce very rapidly, forming 'blooms' and colouring the water because of their abundance. Other species occur on damp soil, dripping ledges, cliffs, rocks above the tide level, and beneath the soil to a depth of one metre or more. Members of the Nostocaceae (*Anabaena, Nostoc*) have the ability to fix atmospheric nitrogen, and many species are the algal partners in lichens (*Nostoc, Stigonema*) or form symbiotic associations with higher plants (*Anabaena* in roots of *Cycas*).

Mostly filamentous, but unicellular and colonial forms exist.
Unicellular forms arise as a result of the separation of the daughter cells after division. Non-filamentous colonies occur when daughter cells remain together after division, held by gelatinous envelopes. Filamentous forms develop following repeated divisions in a single plane. A single row of cells is called a trichome; a trichome plus a gelatinous sheath is termed a filament. Most trichomes are unbranched.

Procaryotic cells.
The cell wall consists of an inner portion of cellulose and an outer gelatinous portion. The protoplasm is differentiated into two regions; the outer region, the chromoplasm, contains photosynthetic pigments and granules, and the inner region is thought to contain the chromatin material. There are no internal cell membranes or endoplasmic reticulum.

Flagella absent.
Only members of the Oscillatoriaceae show any spontaneous movement: a gliding motion possibly caused by the secretion of gelatinous material through minute pores in the cell wall, or rhythmic waves of alternate expansion and contraction along the length of a trichome.

Photosynthetic pigments include chlorophyll *a*, carotenes, xanthophylls, and the biloproteins, C-phycocyanin and C-phycoerythrin.
Variations in the proportions of red, blue and yellow pigments give rise to many different colours in this division. Colour may also be due to non-photosynthetic pigments in the gelatinous envelope.

Storage products are cyanophycin (protein), oil and cyanophycean starch, which is rather similar in structure to glycogen.

Asexual reproduction takes place by means of different types of spores derived from vegetative cells.
Most of the filamentous genera produce akinetes, which are non-motile cells with thick walls and accumulated food reserves. They are often, though not always, formed at specific sites along a trichome. They probably function as resting spores, enabling the alga to survive unfavourable circumstances and giving rise to vegetative filaments when favourable conditions return.

In some genera, endospore formation occurs. During this process, a vegetative cell enlarges and the protoplast divides up to form a number of small spores, similar to the formation of non-motile spores in the Chlorophyta.

Heterocyst formation is common in filamentous genera. The heterocysts develop from enlarged vegetative cells, which gradually lose their contents, and may be terminal or intercalary in position along the filament. Their function is rather obscure, but it seems likely that they are associated with the fragmentation of trichomes, and they often appear next to akinetes. It has also been suggested that heterocysts are associated with the fixation of atmospheric nitrogen.

Hormogonia, consisting of short lengths of trichomes which become detached from the main filament, are found in several genera. In some cases, hormogonia developing at the tips of trichomes become multicellular, spore-like bodies within very thick walls. These are called hormospores.

No sexual reproduction has been observed, but recent work suggests that there is a possibility of interchange of genetic material in a manner similar to that shown by bacteria.

The Cyanophyta are considered to be a very primitive group of organisms, dating back to pre-Cambrian times. Today, there are about 150 genera containing some 1500 species.

Nostoc

DIVISION CYANOPHYTA

CLASS CYANOPHYCEAE

ORDER OSCILLATORIALES

(HORMOGONALES)

GENUS *Nostoc*

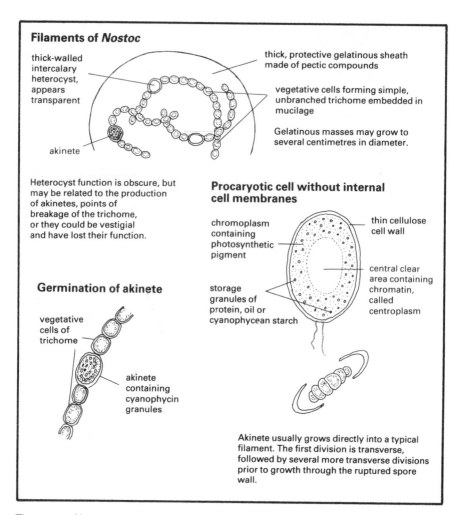

Filaments of *Nostoc*

thick-walled intercalary heterocyst, appears transparent

thick, protective gelatinous sheath made of pectic compounds

vegetative cells forming simple, unbranched trichome embedded in mucilage

Gelatinous masses may grow to several centimetres in diameter.

akinete

Heterocyst function is obscure, but may be related to the production of akinetes, points of breakage of the trichome, or they could be vestigial and have lost their function.

Procaryotic cell without internal cell membranes

chromoplasm containing photosynthetic pigment

thin cellulose cell wall

storage granules of protein, oil or cyanophycean starch

central clear area containing chromatin, called centroplasm

Germination of akinete

vegetative cells of trichome

akinete containing cyanophycin granules

Akinete usually grows directly into a typical filament. The first division is transverse, followed by several more transverse divisions prior to growth through the ruptured spore wall.

The genus *Nostoc* has a world-wide distribution; it is commonly found in soils, on mud, rocks or other plants, and forms the algal partner, or phycobiont, in a number of lichens, e.g. *Peltigera polydactyla*. It can fix atmospheric nitrogen, if molybdenum is present, indicating possibly that the fixation follows a similar pathway to that found in the nitrogen-fixing bacteria. *Nostoc* forms symbiotic associations with higher plants and has been identified in *Anthoceros, Cycas* and some flowering plants.

Nostoc has been used by man as a binding agent on soils, particularly in India, where it adds to the organic matter and increases the nitrogen content of the soil. It has been used as food in China, but, apart from its high nitrogen content, its value there may be in supplying roughage to a diet largely made up of rice and fish.

Division Chlorophyta
Green algae

Mostly freshwater, many terrestrial and a few marine species.
About 90 per cent of the species in this division are freshwater. The terrestrial forms occur on rocks, in soil, on damp wood or bark; there are one or two parasitic species. Some genera form symbiotic associations with animals (zoochlorellae in the endoderm cells of *Hydra*) and others are the algal partners in lichens (*Chlorella, Trebouxia*). The marine species are found in the intertidal zone, growing close to the shore.

Wide range of vegetative structures from unicellular, colonial, coenobial, filamentous, thalloid to siphonaceous.

Eucaryotic cells, with cellulose cell walls and a distinct nucleus containing one or more nucleoli.
Motile cells have a stigma, or eye-spot, and in some genera contractile vacuoles occur.

Flagella present in some genera, normally 2 or 4, acronematic, inserted anteriorly.
Flagella are present in the vegetative cells of the Volvocales and in the reproductive cells of most other orders.

Photosynthetic pigments include chlorophylls *a* and *b*, carotenes and xanthophylls, but no biloproteins.

Chromatophores (chloroplasts): one or more, with characteristic shape for a particular genus or species.
Chloroplasts show many different forms; they can be cup-shaped, stellate, parietal, ribbon-shaped, discoid. They almost always have pyrenoids associated with them. A pyrenoid is a proteinaceous structure surrounded by a sheath of starch grains and thought to be involved in starch formation.

Storage product is starch.

Asexual reproduction is most commonly by zoospores or aplanospores.
These spores are usually formed in vegetative cells, very rarely in special sporangia. Zoospores may be formed singly or in multiples of 2; they are usually released through a pore in the wall of the parent cell and are naked (have no cell wall) during their motile period. When they come to rest, their flagella are either withdrawn or lost, and the protoplast secretes a cellulose cell wall. Aplanospores are non-flagellated and always have a definite cell wall. If the wall is greatly thickened, these spores are referred to as hypnospores.
In some filamentous forms, fragmentation occurs when the filaments break up into sections, each consisting of a few cells.

Sexual reproduction is widespread and either isogamous, anisogamous or oogamous.
The simplest form of sexual reproduction is isogamy, where fusion of flagellated gametes of the same size occurs. In anisogamy, both gametes are

flagellated, but one of the pair is usually considerably larger than the other. In oogamy, there is union of a small, flagellated male gamete (called an antherozoid) with a larger, non-motile, non-flagellated female gamete (called an egg or oosphere). Certain species are homothallic, the gametes from a single parent cell uniting, whereas others are heterothallic, and fusion only occurs between gametes derived from different parent cells.

Zygotes are of two types: thin-walled, germinating quickly, and thick-walled, remaining dormant for a time.

Vegetative cells in this division are normally haploid, meiosis occurring after fusion of the gametes and before germination of the zygote.

Some genera show a well-marked alternation of haploid and diploid generations, which may have morphological differences.

Usually two classes are recognised in this division; the Chlorophyceae containing 425 genera, arranged in 14 orders, with some 6500 species, and the Charophyceae, the stoneworts, containing 6 living genera and about 250 living species.

Class Chlorophyceae

This class includes all the green algae, with the exception of the stoneworts. The Chlorophyceae may be unicellular or multicellular. They do not have ensheathing structures around the sex organs, nor any differentiation of the multicellular, filamentous thallus into nodes and internodes.

Examples:

Order Volvocales	*Chlamydomonas*
	Volvox
Order Chlorococcales	*Chlorella*
Order Ulvales	*Ulva lactuca*
	Enteromorpha spp.
Order Conjugales	*Spirogyra*
Order Chaetophorales	*Pleurococcus*
Order Caulerpales	*Codium*

Order Volvocales

Almost all the genera in this order are found in freshwater and are most abundant where there is a high soluble nitrogen content.

Unicellular or colonial, motile.

In most genera the cells are biflagellate, the flagella being of equal length. Movement is achieved by waves passing from base to tip of the flagellum.

Chloroplasts are usually cup-shaped; one or two genera are colourless and in some cases there is no cellulose cell wall. Pyrenoids are present, commonly one per cell.

Asexual reproduction in the unicellular genera involves division of the cell contents into a definite number of zoospores, which are then released from the parent cell.

In the colonial genera, each colony consists of a definite number of cells arranged in a constant way and should correctly be referred to as a coenobium, e.g. *Pandorina* spp. 4 to 32 cells, *Volvox* spp. 500 to 60 000 cells. Only certain cells will undergo division to form new daughter coenobia.

Sexual reproduction in the unicellular genera begins with the division of a vegetative cell to form naked gametes, which then undergo isogamy or anisogamy depending on the species.

In the colonial genera, the situation varies; in *Gonium*, all the cells divide to form gametes; in *Eudorina*, only certain cells form gametes; in *Volvox*, there is a very definite oogamy with motile male gametes and non-motile female gametes.

DIVISION CHLOROPHYTA

CLASS CHLOROPHYCEAE

ORDER VOLVOCALES

GENUS *Chlamydomonas*

Chlamydomonas

Vegetative cell of *Chlamydomonas*

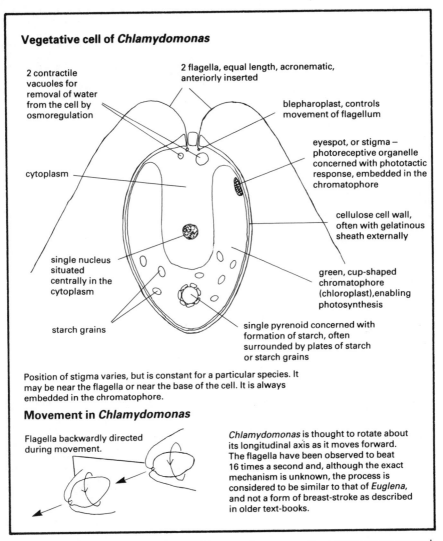

2 contractile vacuoles for removal of water from the cell by osmoregulation

2 flagella, equal length, acronematic, anteriorly inserted

blepharoplast, controls movement of flagellum

eyespot, or stigma – photoreceptive organelle concerned with phototactic response, embedded in the chromatophore

cytoplasm

cellulose cell wall, often with gelatinous sheath externally

single nucleus situated centrally in the cytoplasm

green, cup-shaped chromatophore (chloroplast), enabling photosynthesis

starch grains

single pyrenoid concerned with formation of starch, often surrounded by plates of starch or starch grains

Position of stigma varies, but is constant for a particular species. It may be near the flagella or near the base of the cell. It is always embedded in the chromatophore.

Movement in *Chlamydomonas*

Flagella backwardly directed during movement.

Chlamydomonas is thought to rotate about its longitudinal axis as it moves forward. The flagella have been observed to beat 16 times a second and, although the exact mechanism is unknown, the process is considered to be similar to that of *Euglena*, and not a form of breast-stroke as described in older text-books.

Chlamydomonas species are very widespread; one species grows on snow and ice (*C. nivalis*: it gives a red colour to snow because of the accumulation of carotenoid pigments in its cytoplasm). Other species grow on water-logged patches of soil, acid mine wastes, epiphytically on Bryophyta and on the pileus of some Basidiomycetes, and in drinking water reservoirs, where they can cause problems by clogging up filters.

Chlamydomonas has also been used in the study of metabolic pathways, because it can readily be induced to form biochemical mutants.

Chlamydomonas

Asexual cycle in *Chlamydomonas*

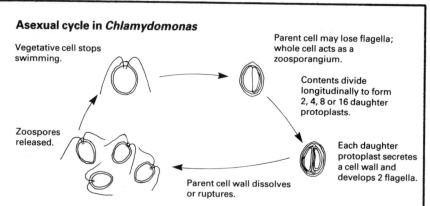

Vegetative cell stops swimming.

Parent cell may lose flagella; whole cell acts as a zoosporangium.

Contents divide longitudinally to form 2, 4, 8 or 16 daughter protoplasts.

Zoospores released.

Each daughter protoplast secretes a cell wall and develops 2 flagella.

Parent cell wall dissolves or ruptures.

Sometimes the protoplast of a vegetative cell may round up, develop a thick wall and become a hypnospore.

Occasionally, when water is in short supply, *Chlamydomonas* spp. growing on soil may form 'palmella' stages, when daughter cells fail to develop flagella after division and remain within the matrix formed when the parent cell wall breaks down. Further divisions may occur, resulting in an irregular colony of cells embedded in a gelatinous matrix. When more water is available, the cells quickly develop flagella and move away.

Sexual cycle in *Chlamydomonas*

Variations of this process occur in different species, e.g. division of contents of vegetative cell may take place to produce gametes. In heterothallic species, when gametes of compatible strains mix, clumping occurs first, followed by pairing.

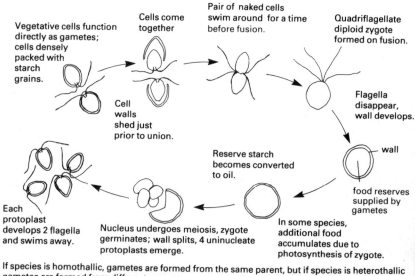

Vegetative cells function directly as gametes; cells densely packed with starch grains.

Cells come together

Pair of naked cells swim around for a time before fusion.

Quadriflagellate diploid zygote formed on fusion.

Cell walls shed just prior to union.

Flagella disappear, wall develops.

Reserve starch becomes converted to oil.

wall

food reserves supplied by gametes

Each protoplast develops 2 flagella and swims away.

Nucleus undergoes meiosis, zygote germinates; wall splits, 4 uninucleate protoplasts emerge.

In some species, additional food accumulates due to photosynthesis of zygote.

If species is homothallic, gametes are formed from the same parent, but if species is heterothallic gametes are formed from different parents.

DIVISION CHLOROPHYTA

CLASS CHLOROPHYCEAE

ORDER VOLVOCALES

GENUS *Volvox*

Volvox

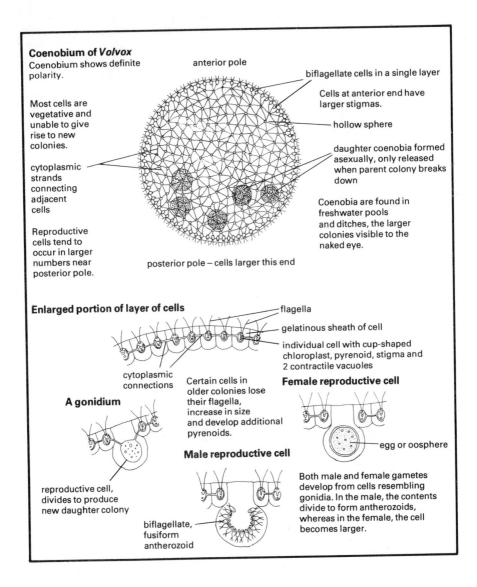

Coenobium of *Volvox*

Coenobium shows definite polarity.

Most cells are vegetative and unable to give rise to new colonies.

cytoplasmic strands connecting adjacent cells

Reproductive cells tend to occur in larger numbers near posterior pole.

anterior pole

biflagellate cells in a single layer

Cells at anterior end have larger stigmas.

hollow sphere

daughter coenobia formed asexually, only released when parent colony breaks down

Coenobia are found in freshwater pools and ditches, the larger colonies visible to the naked eye.

posterior pole – cells larger this end

Enlarged portion of layer of cells

flagella

gelatinous sheath of cell

individual cell with cup-shaped chloroplast, pyrenoid, stigma and 2 contractile vacuoles

cytoplasmic connections

Certain cells in older colonies lose their flagella, increase in size and develop additional pyrenoids.

A gonidium

reproductive cell, divides to produce new daughter colony

Male reproductive cell

biflagellate, fusiform antherozoid

Female reproductive cell

egg or oosphere

Both male and female gametes develop from cells resembling gonidia. In the male, the contents divide to form antherozoids, whereas in the female, the cell becomes larger.

19

Volvox

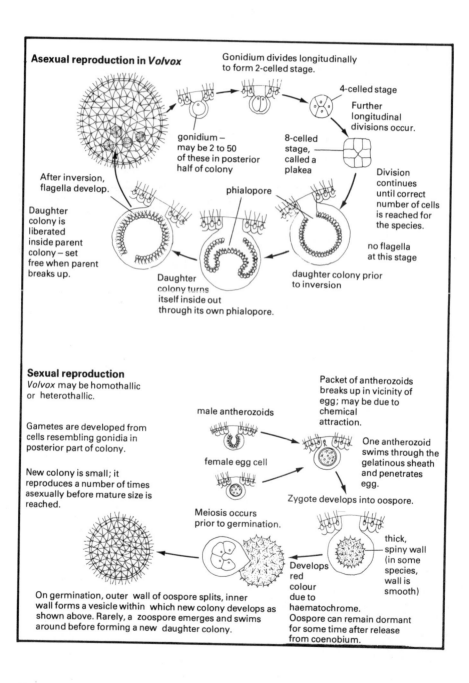

Asexual reproduction in *Volvox*

Gonidium divides longitudinally to form 2-celled stage.

4-celled stage

Further longitudinal divisions occur.

gonidium – may be 2 to 50 of these in posterior half of colony

8-celled stage, called a plakea

Division continues until correct number of cells is reached for the species.

After inversion, flagella develop.

phialopore

no flagella at this stage

Daughter colony is liberated inside parent colony – set free when parent breaks up.

Daughter colony turns itself inside out through its own phialopore.

daughter colony prior to inversion

Sexual reproduction

Volvox may be homothallic or heterothallic.

Gametes are developed from cells resembling gonidia in posterior part of colony.

New colony is small; it reproduces a number of times asexually before mature size is reached.

Packet of antherozoids breaks up in vicinity of egg; may be due to chemical attraction.

male antherozoids

One antherozoid swims through the gelatinous sheath and penetrates egg.

female egg cell

Zygote develops into oospore.

Meiosis occurs prior to germination.

thick, spiny wall (in some species, wall is smooth)

Develops red colour due to haematochrome.

On germination, outer wall of oospore splits, inner wall forms a vesicle within which new colony develops as shown above. Rarely, a zoospore emerges and swims around before forming a new daughter colony.

Oospore can remain dormant for some time after release from coenobium.

Order Chlorococcales

All freshwater genera, many occurring in the plankton of ponds and lakes.

Unicellular or colonial, not filamentous. Colonies may have definite or indefinite numbers of cells.

Normal vegetative cells are non-motile, but biflagellated, ellipsoidal zoospores are formed during asexual reproduction in some genera. Cells may be uninucleate or multinucleate. Unicellular forms bears a close resemblance to *Chlamydomonas*, except that they do not have flagella, stigmas or contractile vacuoles.

Asexual reproduction in the unicellular genera involves the division of the cell contents to form 8 or 16 naked, biflagellated zoospores, all the nuclear divisions occurring first, followed by cleavage of the cytoplasm. Some species, e.g. *Chlorella*, do not form zoospores.

Sexual reproduction has been described for some genera, but has not been recorded for *Chlorella*.

Colonial genera consist either of simple aggregations of cells in mucilage, or more regular, well-defined coenobia, e.g. *Scenedesmus* (4 to 8 cells), *Pediastrum* (a flat plate of from 2 to 128 cells) and *Hydrodictyon* (an open network of cylindrical cells).

Chlorella

Vegetative cells of *Chlorella*

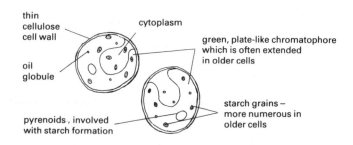

thin cellulose cell wall

cytoplasm

green, plate-like chromatophore which is often extended in older cells

oil globule

starch grains – more numerous in older cells

pyrenoids, involved with starch formation

Asexual reproduction – no sexual stages have been observed in *Chlorella*

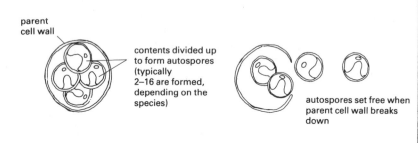

parent cell wall

contents divided up to form autospores (typically 2–16 are formed, depending on the species)

autospores set free when parent cell wall breaks down

All nuclear divisions occur first, followed by cleavage of the cytoplasm.

The genus *Chlorella*, although consisting of very simple organisms, is of great interest and importance. Species are commonly found in small pools and ponds, particularly where the pH is neutral to alkaline. Some species have been recorded from snow and ice, some from the pileus of Basidiomycetes and others live symbiotically with plants and animals. The minute green cells of the zoochlorellae of *Hydra* have been identified by some as members of this genus, and genera closely related to *Chlorella* are common phycobionts in lichens.

Because it is easy to grow in culture, *Chlorella* is a valuable tool for the biochemist and was used to investigate the mechanism of photosynthesis, as well as other metabolic pathways.

Chlorella is also used in the purification of sewage effluents; when grown in shallow tanks, it supplies oxygen for bacterial action and takes up the mineral salts released in the breakdown of the sewage.

Much interest has been shown in growing *Chlorella* for food. It is an efficient converter of light energy and produces a great deal of high-grade protein. However, although production is not difficult, it has been costed as more expensive than growing and refining soya protein.

Order Ulvales

The majority of the genera in this order are marine, though some may be found in brackish or fresh water. They grow profusely in waters that are polluted by sewage.

The multicellular plant body consists of uninucleate cells that divide in two or three planes to produce a parenchymatous structure, which can be a sheet-like thallus, a hollow tube or a solid cylinder.

The reproductive stages are flagellate; in asexual reproduction quadriflagellate zoospores are produced, and in sexual reproduction the gametes are biflagellate.

Each cell of the thallus contains a green, cup-shaped chloroplast and a single pyrenoid, and is capable of division.

Vegetative propagation may occur through fragmentation of the thallus.

Isomorphic alternation of generations occurs; there is a diploid sporophyte stage which undergoes meiosis at zoospore formation, followed by a haploid gametophyte stage. All the vegetative plants look identical and can only be distinguished when reproducing.

Ulva lactuca

DIVISION CHLOROPHYTA

CLASS CHLOROPHYCEAE

ORDER ULVALES

GENUS *Ulva*

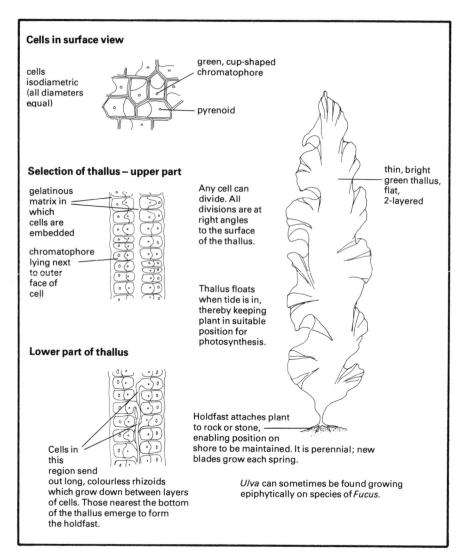

Cells in surface view

cells isodiametric (all diameters equal)

green, cup-shaped chromatophore

pyrenoid

Selection of thallus – upper part

gelatinous matrix in which cells are embedded

chromatophore lying next to outer face of cell

Any cell can divide. All divisions are at right angles to the surface of the thallus.

Thallus floats when tide is in, thereby keeping plant in suitable position for photosynthesis.

Lower part of thallus

Cells in this region send out long, colourless rhizoids which grow down between layers of cells. Those nearest the bottom of the thallus emerge to form the holdfast.

thin, bright green thallus, flat, 2-layered

Holdfast attaches plant to rock or stone, enabling position on shore to be maintained. It is perennial; new blades grow each spring.

Ulva can sometimes be found growing epiphytically on species of *Fucus*.

Ulva lactuca is a common marine alga, found in rock pools in the mid-tidal region of the shore. It can tolerate highly alkaline conditions, and has been recorded in rock pools where the pH reached 10.

Ulva lactuca, the sea lettuce, is edible, and though not as good as the purple laver, *Porphyra*, for flavour, it has been used as a substitute. It is collected by fishermen round the coasts of China, and sold for food or used in the production of a medicine to combat fever.

Ulva lactuca

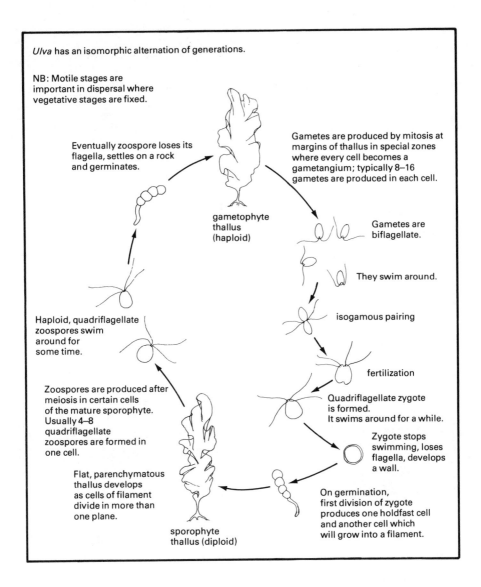

Ulva has an isomorphic alternation of generations.

NB: Motile stages are important in dispersal where vegetative stages are fixed.

Eventually zoospore loses its flagella, settles on a rock and germinates.

gametophyte thallus (haploid)

Gametes are produced by mitosis at margins of thallus in special zones where every cell becomes a gametangium; typically 8–16 gametes are produced in each cell.

Gametes are biflagellate.

They swim around.

Haploid, quadriflagellate zoospores swim around for some time.

isogamous pairing

fertilization

Zoospores are produced after meiosis in certain cells of the mature sporophyte. Usually 4–8 quadriflagellate zoospores are formed in one cell.

Quadriflagellate zygote is formed. It swims around for a while.

Zygote stops swimming, loses flagella, develops a wall.

Flat, parenchymatous thallus develops as cells of filament divide in more than one plane.

On germination, first division of zygote produces one holdfast cell and another cell which will grow into a filament.

sporophyte thallus (diploid)

Enteromorpha

DIVISION CHLOROPHYTA

CLASS CHLOROPHYCEAE

ORDER ULVALES

GENUS *Enteromorpha*

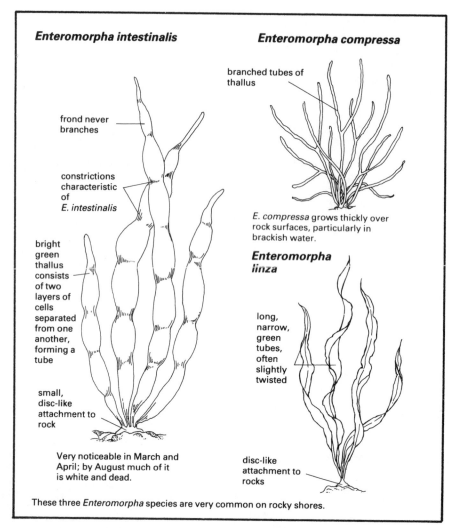

Enteromorpha intestinalis

frond never branches

constrictions characteristic of *E. intestinalis*

bright green thallus consists of two layers of cells separated from one another, forming a tube

small, disc-like attachment to rock

Very noticeable in March and April; by August much of it is white and dead.

Enteromorpha compressa

branched tubes of thallus

E. compressa grows thickly over rock surfaces, particularly in brackish water.

Enteromorpha iinza

long, narrow, green tubes, often slightly twisted

disc-like attachment to rocks

These three *Enteromorpha* species are very common on rocky shores.

Enteromorpha species are common in the upper parts of the intertidal zone, particularly where freshwater streams flow down into the sea. In such areas, they often completely cover the rocks, providing shelter for the young plants of *Fucus* spp. If, however, there is a large population of limpets (*Patella* sp.), most of the *Enteromorpha* gets eaten, with the result that it is difficult for the *Fucus* plants to become established. Like *Ulva*, *Enteromorpha* has an isomorphic alternation of generations, but the gametes vary in size; the female gametes are large and bright green, and the male gametes are smaller, narrow and yellowish-green.

Order Conjugales

All the genera in this order are found in freshwater.

Unicellular, or filamentous. Cylindrical cells are joined end to end to form unbranched filaments.

Flagellate stages absent.

The chloroplasts can be quite elaborate; plate-like, helical or stellate depending on the genus. The cell walls have an inner cellulose layer and an outer pectic layer. Usually the cells possess large vacuoles with the nucleus suspended in the centre by cytoplasmic strands.

In filamentous genera, vegetative propagation may occur by fragmentation.

There is no asexual spore production in this order.

Sexual reproduction, called conjugation, is achieved by the fusion of amoeboid aplanogametes, formed singly within the cells. The union of the gametes is brought about by means of tubular connections, or conjugation tubes, between two cells. The zygote formed becomes a zygospore, with a thick wall and food reserves. It can undergo a prolonged period of dormancy during which dispersal can occur.

Spirogyra

DIVISION CHLOROPHYTA

CLASS CHLOROPHYCEAE

ORDER CONJUGALES

GENUS *Spirogyra*

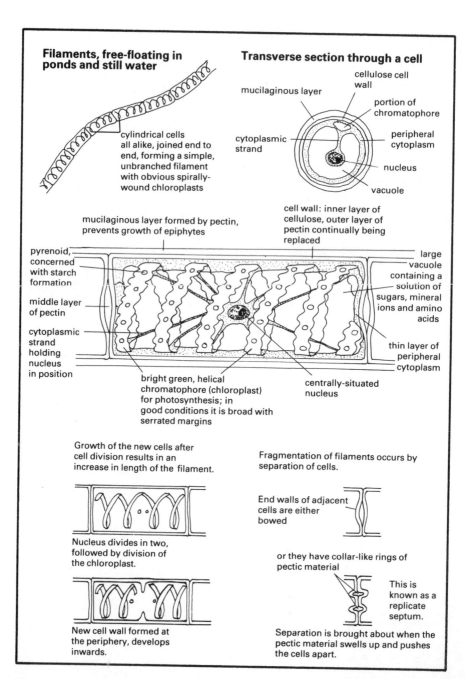

Filaments, free-floating in ponds and still water

cylindrical cells all alike, joined end to end, forming a simple, unbranched filament with obvious spirally-wound chloroplasts

Transverse section through a cell

cellulose cell wall

mucilaginous layer

portion of chromatophore

cytoplasmic strand

peripheral cytoplasm

nucleus

vacuole

mucilaginous layer formed by pectin, prevents growth of epiphytes

cell wall: inner layer of cellulose, outer layer of pectin continually being replaced

pyrenoid, concerned with starch formation

large vacuole containing a solution of sugars, mineral ions and amino acids

middle layer of pectin

cytoplasmic strand holding nucleus in position

thin layer of peripheral cytoplasm

bright green, helical chromatophore (chloroplast) for photosynthesis; in good conditions it is broad with serrated margins

centrally-situated nucleus

Growth of the new cells after cell division results in an increase in length of the filament.

Fragmentation of filaments occurs by separation of cells.

End walls of adjacent cells are either bowed

Nucleus divides in two, followed by division of the chloroplast.

or they have collar-like rings of pectic material

This is known as a replicate septum.

New cell wall formed at the periphery, develops inwards.

Separation is brought about when the pectic material swells up and pushes the cells apart.

DIVISION CHLOROPHYTA
CLASS CHLOROPHYCEAE
ORDER CONJUGALES
GENUS *Spirogyra*

Spirogyra

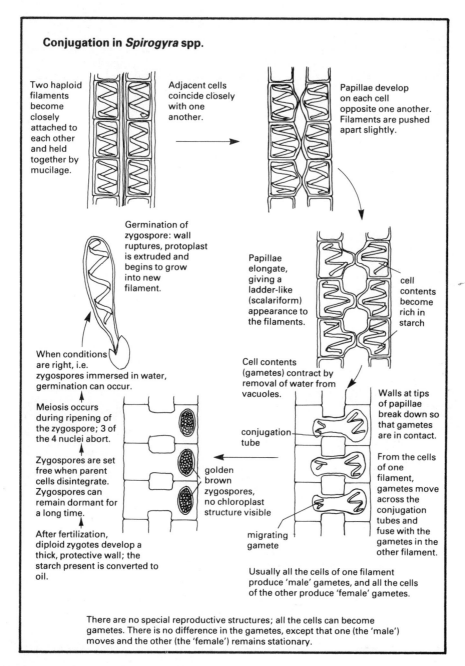

Conjugation in *Spirogyra* spp.

Two haploid filaments become closely attached to each other and held together by mucilage.

Adjacent cells coincide closely with one another.

Papillae develop on each cell opposite one another. Filaments are pushed apart slightly.

Germination of zygospore: wall ruptures, protoplast is extruded and begins to grow into new filament.

Papillae elongate, giving a ladder-like (scalariform) appearance to the filaments.

cell contents become rich in starch

When conditions are right, i.e. zygospores immersed in water, germination can occur.

Cell contents (gametes) contract by removal of water from vacuoles.

Walls at tips of papillae break down so that gametes are in contact.

Meiosis occurs during ripening of the zygospore; 3 of the 4 nuclei abort.

conjugation tube

Zygospores are set free when parent cells disintegrate. Zygospores can remain dormant for a long time.

golden brown zygospores, no chloroplast structure visible

From the cells of one filament, gametes move across the conjugation tubes and fuse with the gametes in the other filament.

After fertilization, diploid zygotes develop a thick, protective wall; the starch present is converted to oil.

migrating gamete

Usually all the cells of one filament produce 'male' gametes, and all the cells of the other produce 'female' gametes.

There are no special reproductive structures; all the cells can become gametes. There is no difference in the gametes, except that one (the 'male') moves and the other (the 'female') remains stationary.

Order Chaetophorales

Almost all the genera in this order are freshwater.

Typically, a branching, filamentous thallus, heterotrichous with a prostrate and an erect system is seen. There is a great deal of variation in this order, and in many genera, either the prostrate or the erect system is reduced, making the heterotrichous nature obscure. In some cases, the filaments of the prostrate system may be packed together to form a pseudo-parenchymatous structure, particularly if the erect system is either absent or reduced. The heterotrichous habit is interesting in that it appears to be the highest level of organization achieved by filamentous algae.

Flagellate cells are produced during asexual and sexual reproduction.

In asexual reproduction, quadriflagellate zoospores are produced, either in normal vegetative cells or in specialized cells which function as zoosporangia.

Sexual reproduction in most genera is isogamous, fusion occurring between bi- or quadri-flagellate gametes, but anisogamy and oogamy may be found.

The genus *Pleurococcus* is placed in this order as it is considered to be a very reduced form and to have evolved from a heterotrichous ancestor. It does not produce any motile zoospores or gametes, but it will form profusely branched filaments when submerged in water.

Much more typical of this order is the genus *Stigeoclonium*, a fairly common alga found growing on submerged woodwork and stones in freshwater.

| DIVISION CHLOROPHYTA |
| CLASS CHLOROPHYCEAE |
| ORDER CHAETOPHORALES |
| GENUS *Pleurococcus* |

Powdery covering on tree-trunks made up of solitary cells, together with 2- and 4-celled colonies.

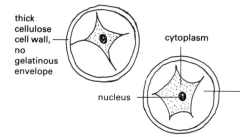

thick
cellulose
cell wall,
no
gelatinous
envelope

cytoplasm

nucleus

Solitary cells are usually
spherical.

single, lobed, parietal, green
chromatophore – usually lacking
pyrenoids

Under very moist conditions, or when kept in culture,
Pleurococcus will form short
filaments.

Vegetative propagation is by simple division, and is frequent when
conditions are favourable, enabling the alga to cover large areas
rapidly.

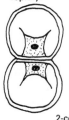

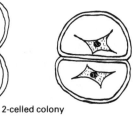

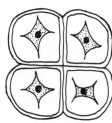

2-celled colony

4-celled colony

After division, the daughter cells may separate, or they may remain
together to form a 2-celled colony; a further division at right angles
to the first results in the formation of a 4-celled colony.

Pleurococcus is one of the commonest green algae in the world. It is frequently
found as a powdery, green covering on the trunks of trees, on stone or brick
walls, and on garden fences and seats. It has been recorded growing on the
bark of trees up to 10 metres above the ground, and it depends directly on rain
or high humidity for its water supply. It can tolerate prolonged periods of
drought, growth taking place when sufficient water is available.

Pleurococcus is often found in greater abundance on the northern and
eastern sides of tree-trunks, and an interesting survey can be carried out,
relating the distribution of this alga to water availability and light intensity.

Order Caulerpales
(Siphonales)

All members of this order are marine and are most frequently found in tropical and sub-tropical seas.

The thallus is coenocytic, consisting of a single, multinucleate cell, which forms a branched tube capable of indefinite growth. In some genera, the branches are arranged on a definite axis or inter-twined to form large, complex structures.

Some members of the order produce flagellate zoospores and gametes.

Photosynthetic pigments include two xanthophylls not found in any other orders of the Chlorophyceae. The vegetative cells have diploid nuclei, large numbers of small, green, discoid chromatophores and often pyrenoids.

Vegetative propagation occurs in many genera and involves fragmentation of the thallus.

Asexual reproduction by spores is rare, but when it does occur, either flagellate zoospores or aplanospores are produced.

Sexual reproduction is more general and may be isogamous, anisogamous or oogamous, the gametes in some genera being produced in specialized gametangia.

DIVISION CHLOROPHYTA
CLASS CHLOROPHYCEAE
ORDER CAULERPALES (SIPHONALES)
GENUS *Codium*

Codium tomentosum

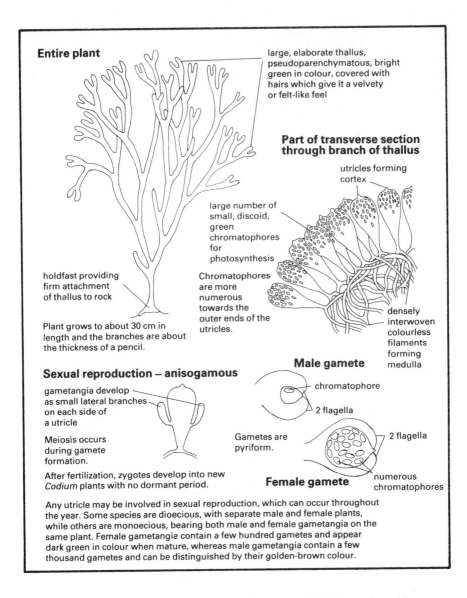

Entire plant

large, elaborate thallus, pseudoparenchymatous, bright green in colour, covered with hairs which give it a velvety or felt-like feel

Part of transverse section through branch of thallus

utricles forming cortex

large number of small, discoid, green chromatophores for photosynthesis

holdfast providing firm attachment of thallus to rock

Chromatophores are more numerous towards the outer ends of the utricles.

densely interwoven colourless filaments forming medulla

Plant grows to about 30 cm in length and the branches are about the thickness of a pencil.

Sexual reproduction – anisogamous

Male gamete

gametangia develop as small lateral branches on each side of a utricle

chromatophore

2 flagella

Meiosis occurs during gamete formation.

Gametes are pyriform.

2 flagella

After fertilization, zygotes develop into new *Codium* plants with no dormant period.

Female gamete

numerous chromatophores

Any utricle may be involved in sexual reproduction, which can occur throughout the year. Some species are dioecious, with separate male and female plants, while others are monoecious, bearing both male and female gametangia on the same plant. Female gametangia contain a few hundred gametes and appear dark green in colour when mature, whereas male gametangia contain a few thousand gametes and can be distinguished by their golden-brown colour.

The *Codium* genus contains about 50 species, most of which are found in tropical waters, but *Codium tomentosum* is fairly common in deep pools in the middle of the littoral zone of rocky shores in the south and west of Britain. It forms the food of the sea slug, *Elysia viridis*. The Japanese prepare a food called miru from species of *Codium*.

Class Charophyceae

This class includes the stoneworts, which all show a much more complex morphology and a greater specialization of reproductive processes than the Chlorophyceae. All have an erect thallus, differentiated into a regular arrangement of nodes and internodes, and sterile tissue associated with the sex organs. All members of this class are placed in a single order, and most present-day stoneworts are found in freshwater.

Examples: Order Charales   *Chara*

Order Charales

Most living members of this order are found in alkaline freshwater, frequently covering extensive areas on the muddy or sandy bottom of ponds. Many species, especially *Chara*, become encrusted with calcium carbonate.

The thallus is always multicellular, with a clear main axis attached to the substratum by means of multicellular rhizoids. The main axis is differentiated into nodes, which bear whorls of several branches referred to as 'leaves'. Axillary branches capable of unlimited growth may develop.

Cells are multinucleate, containing large numbers of discoid chloroplasts.

No asexual reproduction by spore formation occurs, but vegetative propagation may occur by the detachment of groups of cells formed as tuber- or protonema-like outgrowths on the nodes, or from bulbils which develop on the rhizoids.

Sexual reproduction is oogamous. Antheridia and oogonia are formed on the nodes of 'leaves' and are surrounded by envelopes of sterile tissue. A mature antheridium is produced after a complicated sequence of divisions. An initial cell divides transversely to form a lower pedicel cell, and an upper cell which undergoes two successive divisions to give four quadrately arranged cells. Each of these cells then divides transversely giving eight cells. Each one of these eight cells undergoes further divisions to produce an outer shield cell, a middle cell called the manubrium and an inner cell which eventually gives rise to antheridial filaments.
   The number of cells in an antheridial filament varies, ranging from 5 to 50 in the same species. Each cell of a filament becomes an antheridium, inside which a single, elongated, biflagellate, spiral antherozoid develops. When the antherozoids are mature, the shield cells separate from each other and expose the antheridial filaments. The antherozoids escape through pores in the walls of the antheridia.
   An oogonium arises from an initial cell which divides to form a row of three cells. The terminal cell of this row, called an oogonial mother cell, elongates and then divides transversely to form an oogonium, containing a single egg, and a short stalk cell. The lower cell forms the pedicel and the middle cell divides to form five sheath initials, which divide transversely forming two tiers of cells. The uppermost tier of cells forms a corona, and the lower ones become elongated and twist round to form a tube, or sheath, around the oogonium. Inside the oogonium, the egg accumulates food reserves and becomes opaque. When the egg cell is mature, the tube cells separate from one another just below the corona and five slits appear, through which the antherozoids swim to bring about fertilization.
   The zygote secretes a thick wall and (after becoming detached from the parent plant) falls to the bottom of the pond, where it undergoes a period of dormancy. The diploid zygote nucleus divides meiotically before germination. A green filament, or protonema, is produced when the zygote germinates, and this differentiates to give the nodes and internodes of the main axis of a new plant.

Chara fragilis

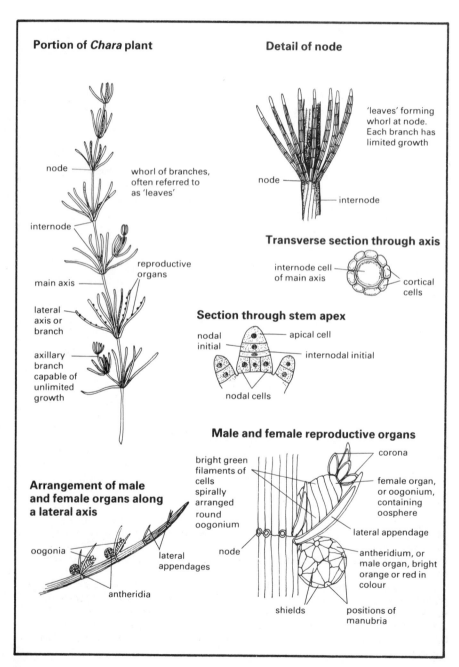

Portion of *Chara* plant

node

whorl of branches, often referred to as 'leaves'

internode

reproductive organs

main axis

lateral axis or branch

axillary branch capable of unlimited growth

Detail of node

'leaves' forming whorl at node. Each branch has limited growth

node

internode

Transverse section through axis

internode cell of main axis

cortical cells

Section through stem apex

nodal initial

apical cell

internodal initial

nodal cells

Male and female reproductive organs

corona

bright green filaments of cells spirally arranged round oogonium

female organ, or oogonium, containing oosphere

lateral appendage

node

antheridium, or male organ, bright orange or red in colour

Arrangement of male and female organs along a lateral axis

oogonia

lateral appendages

antheridia

shields

positions of manubria

Chara gracilis

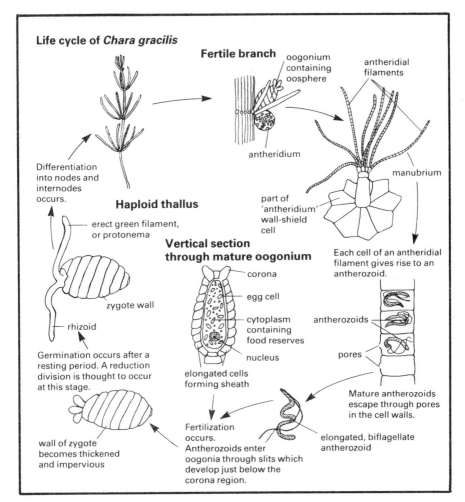

Life cycle of *Chara gracilis*

Fertile branch

oogonium containing oosphere

antheridial filaments

antheridium

Differentiation into nodes and internodes occurs.

manubrium

Haploid thallus

erect green filament, or protonema

part of 'antheridium' wall-shield cell

Vertical section through mature oogonium

zygote wall

corona

egg cell

cytoplasm containing food reserves

nucleus

Each cell of an antheridial filament gives rise to an antherozoid.

antherozoids

rhizoid

pores

Germination occurs after a resting period. A reduction division is thought to occur at this stage.

elongated cells forming sheath

Mature antherozoids escape through pores in the cell walls.

wall of zygote becomes thickened and impervious

Fertilization occurs. Antherozoids enter oogonia through slits which develop just below the corona region.

elongated, biflagellate antherozoid

Chara species grow in still fresh or brackish water in temperate climates, thriving best where the water is clear, hard and unpolluted. There are only 6 living genera in the Order Charales, containing some 250 species. The fossil record extends back to the Devonian period and the forms found in these rocks differ very little from present day types.

Chara gracilis grows to a height of about 30 cm. Growth of the axis is brought about by means of a dome-shaped apical cell, cutting off nodal and internodal initials. The nodal initials give rise to whorls of branches, most of which have limited growth. In all the *Chara* spp. certain cells at the node give rise to filaments which grow round the internode cells, forming a structure resembling a cortex.

Division Euglenophyta
Euglenoids

Mostly freshwater, but some terrestrial, brackish or marine.
Most genera appear to thrive best in stagnant freshwater, which has a high
organic content.

Unicellular flagellates.
No filamentous or true colonial forms occur, although palmelloid or
dendroid colonies are found in one genus (*Colacium*). Cells are elongated
and either radial or flattened.

Eucaryotic cells with no cellulose cell walls.
The protoplast is bounded by a periplast, or pellicle, which may have definite
rigidity and often shows sculpturing.

Flagella: 1–3, typically 2, pantonematic with a row of barbs on one side only.
Flagella arise from the base of the reservoir and pass to the exterior by
means of a narrow passage called the gullet. Movement is achieved by the
passage of waves from the base to the tip of the flagellum. These waves
cause the organism to rotate about its own axis and to move forward in a
helical path. The longer the flagellum, the faster the movement.

Photosynthetic pigments, when present, include chlorophylls *a* and *b*, beta-
carotene and xanthophylls. They are very similar, but not identical, to the
photosynthetic pigments found in the Chlorophyta.

Chromatophores in pigmented species are variable in form: discoid, stellate,
rod-like or ribbon-shaped.

Storage products are a starch-like polysaccharide, paramylum, and oils.
The paramylum may occur as discs, rings, rods or spherical granules in the
cytoplasm.

Asexual reproduction is by longitudinal fission when a limiting size is reached.
The division starts at the anterior end and works backwards. Organelles,
which do not divide, then reform in daughter cells as needed. Some species
may become enclosed in a sheath of mucilage during division, and this can
give rise to a temporary palmelloid condition.

Sexual reproduction has not been demonstrated conclusively for any members
of this division.

In several genera, thick-walled resting stages, or cysts, may be formed when
conditions are unfavourable for growth.

In the past there has been lengthy debate about the classification of the
euglenoids, but nowadays it does seem convenient to include the
photosynthetic species in the algae. Their photosynthetic pigments resemble
those of other plants and are not found in animals. It is interesting to note that
an animal pigment, astaxanthin, has been isolated from the stigma of *Euglena*.

There are about 25 genera and some 450 species. The division is usually split into two orders: Euglenales, containing both pigmented and unpigmented genera, which are phototrophic or saptotrophic; Peranematales, containing unpigmented genera, which are phagotrophic. The latter are best grouped with the Protozoa.

Euglena viridis

DIVISION EUGLENOPHYTA

CLASS EUGLENOPHYCEAE

ORDER EUGLENALES

GENUS *Euglena*

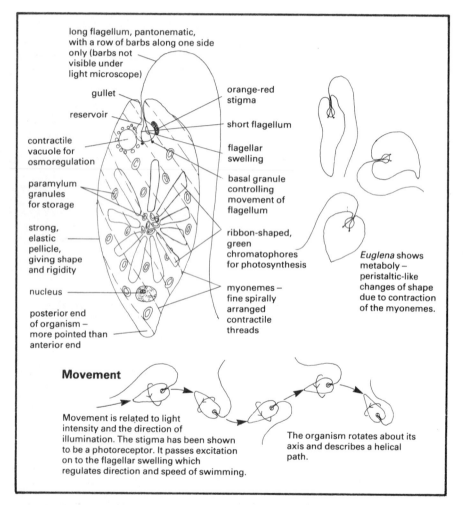

long flagellum, pantonematic, with a row of barbs along one side only (barbs not visible under light microscope)

gullet

reservoir

contractile vacuole for osmoregulation

paramylum granules for storage

strong, elastic pellicle, giving shape and rigidity

nucleus

posterior end of organism – more pointed than anterior end

orange-red stigma

short flagellum

flagellar swelling

basal granule controlling movement of flagellum

ribbon-shaped, green chromatophores for photosynthesis

myonemes – fine spirally arranged contractile threads

Euglena shows metaboly – peristaltic-like changes of shape due to contraction of the myonemes.

Movement

Movement is related to light intensity and the direction of illumination. The stigma has been shown to be a photoreceptor. It passes excitation on to the flagellar swelling which regulates direction and speed of swimming.

The organism rotates about its axis and describes a helical path.

Euglena viridis is often found in puddles, ponds, ditches or stagnant water generally, flourishing where there is plenty of organic matter. Other species of *Euglena* are epiphytic and epipelic, and can cause problems in lakes and reservoirs. When conditions are favourable for growth, rapid reproduction occurs, forming 'blooms' which can clog up water-filtering apparatus.

The genus *Euglena* is remarkable in that many of the pigmented species can live in the dark, if they are provided with a source of energy to compensate for lack of light. Most of these species will retain their chlorophyll under these circumstances, but some become colourless. Photosynthetic species cannot use nitrates as a source of nitrogen, but must have either ammonium salts or organic nitrogen in order to synthesize proteins.

Division Bacillariophyta
Diatoms

Widespread: freshwater, terrestrial or marine.
Marine species are abundant in the phytoplankton, and many are very sensitive to changes in the temperature and salinity of the water. A few marine species are epilithic, or epiphytic on red and brown seaweeds in the intertidal zone. Most freshwater species are found in ponds or lakes, and terrestrial species grow on walls, cliffs, in soil and on the bark of trees.

Unicellular or colonial.
In colonial forms, the cells are sometimes joined together by mucilage to form stellate or filamentous colonies. Sometimes cells are held together by spines on the frustules, or embedded in a mucilaginous envelope.

Eucaryotic cells, with cell walls in two distinct halves.
The cell wall, called the frustule, consists of pectin impregnated with silica. It is unevenly thickened, the thick and thin areas giving characteristic markings which are used in identification. Each half (or valve) of the frustule is called a theca, and the two halves are joined together by connecting, or girdle, bands. The outer half is called the epitheca, and the inner half the hypotheca. In the Centrales, the cells are centric or polygonal, resembling Petri dishes, and in the Pennales, the cells are elongated, resembling date boxes.

Flagella rare.
In the Centrales, spermatozoids are flagellate, with one pantonematic flagellum. No flagella are found in the Pennales, but some species show gliding movements, which appear to be associated with the production of mucilage.

Photosynthetic pigments are chlorophylls *a* and *c*, carotenes and xanthophylls, including fucoxanthin.

Chromatophores are either numerous and discoid, as in the Centrales, or 2 in number, parietal and plate-like, as in the Pennales.

Storage products are oil, volutin and chrysose, a polysaccharide similar to laminarin.

Asexual reproduction is by binary fission.
The plane of division is always parallel to the valve surfaces. The daughter cell which has the hypotheca of the parent cell secretes a new hypotheca, and is therefore slightly smaller than the parent cell. Thus, in a given population, there will tend to be a progressive decrease in average cell size, which is restored after sexual reproduction.

Sexual reproduction involves the fusion of haploid gametes.
The vegetative cells are diploid and gamete formation is preceded by meiosis. In the Pennales, the gametes are amoeboid and isogamy occurs, whereas in the Centrales, the spermatozoids are flagellate and oogamy occurs. After fusion, the zygote increases in size and becomes an auxospore.

This then undergoes two mitoses; in each case only one daughter cell survives. Eventually a vegetative cell is formed, which has the largest dimensions for the particular species.

There are about 170 genera and 5500 species of diatoms, most of which are still living. There is an extensive fossil record, because, under the right conditions, the silica walls persist long after the death of the protoplast. Over millions of years, diatom frustules have accumulated at the bottom of lakes and seas, forming diatomaceous earth. This substance has been quarried and is used commercially as an abrasive in polishes and toothpastes, in the filtration of liquids, as an inert mineral filler in the manufacture of paints and plastics, and as a resistant lining for boilers and blast furnaces. Because of their fineness and uniformity, the markings on diatom frustules have been used since Victorian times to check the resolving power of the objective lenses of light microscopes. From the fossil records, the centric form appears to be more primitive than the pennate form.

Pinnularia viridis

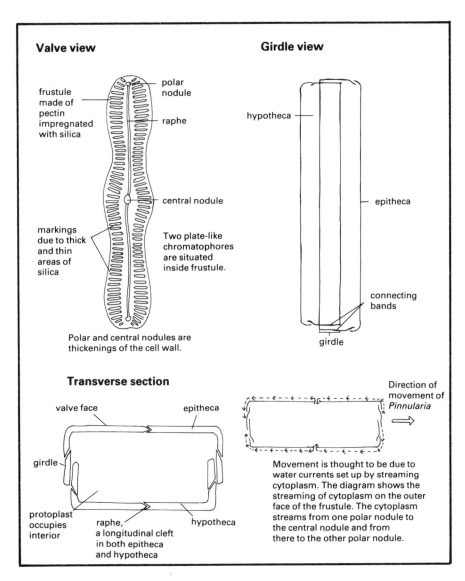

Valve view

frustule made of pectin impregnated with silica

polar nodule

raphe

central nodule

markings due to thick and thin areas of silica

Two plate-like chromatophores are situated inside frustule.

Polar and central nodules are thickenings of the cell wall.

Girdle view

hypotheca

epitheca

connecting bands

girdle

Transverse section

valve face

epitheca

girdle

protoplast occupies interior

raphe, a longitudinal cleft in both epitheca and hypotheca

hypotheca

Direction of movement of *Pinnularia*

Movement is thought to be due to water currents set up by streaming cytoplasm. The diagram shows the streaming of cytoplasm on the outer face of the frustule. The cytoplasm streams from one polar nodule to the central nodule and from there to the other polar nodule.

Members of the genus *Pinnularia* are found in acidic, freshwater pools, or they may be epilithic on acidic rocks, epiphytic on Bryophyta and in moist soils, where they may form the food of soil Protozoa and rotifers.

Pinnularia viridis has been recorded in bog pools where the pH is low, in the region of 3.5 to 4.5, and it can also be found on black, peaty sediments on moorlands.

Pinnularia viridis

DIVISION BACILLARIOPHYTA

CLASS BACILLARIOPHYCEAE

ORDER PENNALES

GENUS *Pinnularia*

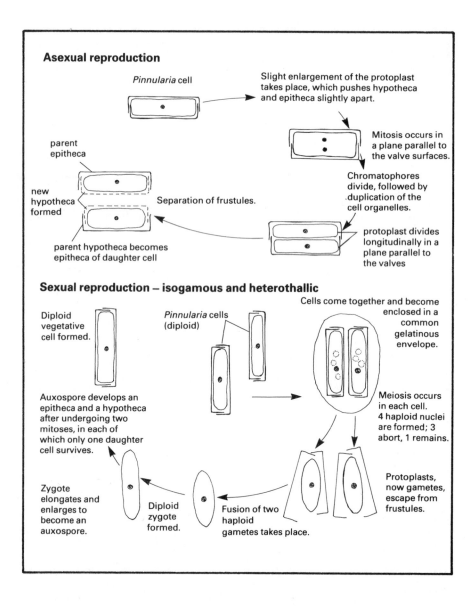

Asexual reproduction

Pinnularia cell

Slight enlargement of the protoplast takes place, which pushes hypotheca and epitheca slightly apart.

Mitosis occurs in a plane parallel to the valve surfaces.

Chromatophores divide, followed by duplication of the cell organelles.

parent epitheca

new hypotheca formed

Separation of frustules.

protoplast divides longitudinally in a plane parallel to the valves

parent hypotheca becomes epitheca of daughter cell

Sexual reproduction – isogamous and heterothallic

Diploid vegetative cell formed.

Pinnularia cells (diploid)

Cells come together and become enclosed in a common gelatinous envelope.

Auxospore develops an epitheca and a hypotheca after undergoing two mitoses, in each of which only one daughter cell survives.

Meiosis occurs in each cell. 4 haploid nuclei are formed; 3 abort, 1 remains.

Zygote elongates and enlarges to become an auxospore.

Diploid zygote formed.

Fusion of two haploid gametes takes place.

Protoplasts, now gametes, escape from frustules.

44

Division Xanthophyta
Yellow-green algae

Mostly freshwater or terrestrial.
> Terrestrial species grow on tree-trunks, damp walls, amongst Bryophyta, or on drying mud.

Unicellular, colonial, filamentous and siphonaceous forms occur.

Eucaryotic cells. Cell walls frequently absent, but, when present, they have a high pectic content and may contain deposits of silica. Cells uninucleate or multinucleate.
> The nuclei are usually very small and not easily seen.

Flagella present, usually 2, unequal, inserted anteriorly; the longer flagellum is pantonematic and the shorter one is acronematic.

Photosynthetic pigments include chlorophylls *a* and *e*, beta-carotene and one or two xanthophylls.
> Chlorophyll *e* has been identified in the zoospores of only 2 genera as yet.

Chromatophores: one or more per cell, discoid, yellow-green and mostly without pyrenoids.
> Where pyrenoids are present, they are not thought to be connected with the accumulation of reserve foods.

Storage products are oil and fat.
> Members of this division do not store starch.

Vegetative multiplication of the filamentous and colonial forms can occur by fragmentation.

Asexual reproduction by means of flagellate zoospores, aplanospores, akinetes or hypnospores.
> Zoospores may be formed singly or in numbers in a cell. They are naked, pyriform, and biflagellate with unequal flagella. They generally have contractile vacuoles and one or more chromatophores. Aplanospores may grow directly into new plants, or they may give rise to zoospores, which will in turn give rise to new plants.

Sexual reproduction may be isogamous, anisogamous or oogamous.

The algae in this division were once included in the Chlorophyceae, but it is now felt that there are sufficient differences to justify the formation of a separate division, with one class, the Xanthophyceae. The class is divided into 6 orders, using vegetative characteristics in the classification, e.g. all the motile forms in one order, all the coccoid forms in another, paralleling the situation found in the Chlorophyceae.
> Example: Order Vaucheriales   *Vaucheria*

Vaucheria

DIVISION XANTHOPHYTA

CLASS XANTHOPHYCEAE

ORDER VAUCHERIALES

GENUS *Vaucheria*

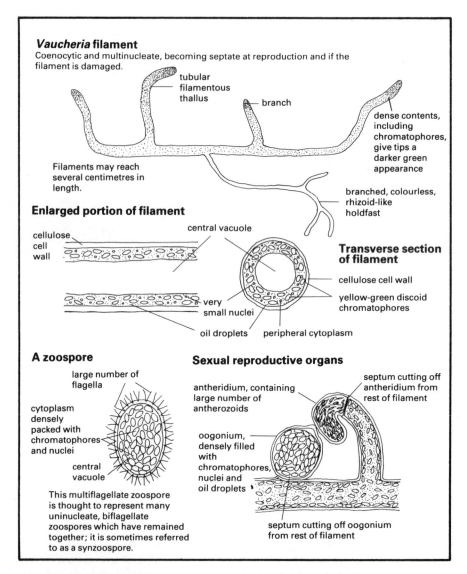

Vaucheria filament
Coenocytic and multinucleate, becoming septate at reproduction and if the filament is damaged.

tubular filamentous thallus

branch

dense contents, including chromatophores, give tips a darker green appearance

Filaments may reach several centimetres in length.

branched, colourless, rhizoid-like holdfast

Enlarged portion of filament

cellulose cell wall

central vacuole

Transverse section of filament

cellulose cell wall

yellow-green discoid chromatophores

very small nuclei

oil droplets   peripheral cytoplasm

A zoospore

large number of flagella

cytoplasm densely packed with chromatophores and nuclei

central vacuole

This multiflagellate zoospore is thought to represent many uninucleate, biflagellate zoospores which have remained together; it is sometimes referred to as a synzoospore.

Sexual reproductive organs

antheridium, containing large number of antherozoids

septum cutting off antheridium from rest of filament

oogonium, densely filled with chromatophores, nuclei and oil droplets

septum cutting off oogonium from rest of filament

Some species of *Vaucheria* are partly terrestrial and form coarse, green masses on damp soil, particularly round the edges of ponds, or on the silty shores of salt-marshes; other species are totally aquatic, mostly in fresh water.

The genus *Vaucheria* has a very elaborate form of oogamous reproduction, which is quite different from that shown by any other members of the Xanthophyceae.

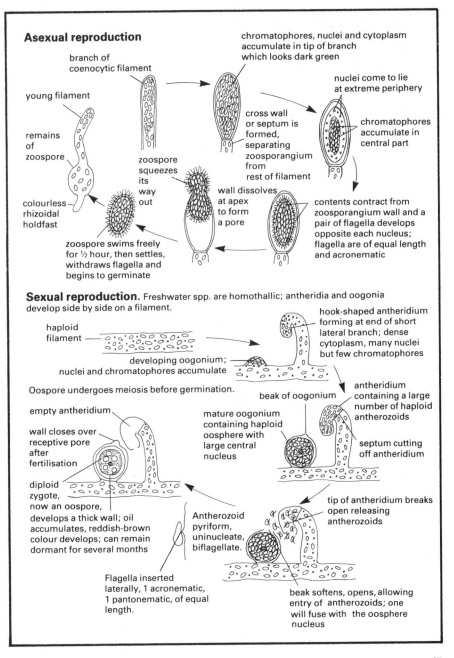

Asexual reproduction

chromatophores, nuclei and cytoplasm accumulate in tip of branch which looks dark green

branch of coenocytic filament

nuclei come to lie at extreme periphery

young filament

cross wall or septum is formed, separating zoosporangium from rest of filament

chromatophores accumulate in central part

remains of zoospore

zoospore squeezes its way out

colourless rhizoidal holdfast

wall dissolves at apex to form a pore

contents contract from zoosporangium wall and a pair of flagella develops opposite each nucleus; flagella are of equal length and acronematic

zoospore swims freely for ½ hour, then settles, withdraws flagella and begins to germinate

Sexual reproduction. Freshwater spp. are homothallic; antheridia and oogonia develop side by side on a filament.

haploid filament

developing oogonium; nuclei and chromatophores accumulate

hook-shaped antheridium forming at end of short lateral branch; dense cytoplasm, many nuclei but few chromatophores

Oospore undergoes meiosis before germination.

beak of oogonium

antheridium containing a large number of haploid antherozoids

empty antheridium

mature oogonium containing haploid oosphere with large central nucleus

wall closes over receptive pore after fertilisation

septum cutting off antheridium

diploid zygote, now an oospore, develops a thick wall; oil accumulates, reddish-brown colour develops; can remain dormant for several months

Antherozoid pyriform, uninucleate, biflagellate.

tip of antheridium breaks open releasing antherozoids

Flagella inserted laterally, 1 acronematic, 1 pantonematic, of equal length.

beak softens, opens, allowing entry of antherozoids; one will fuse with the oosphere nucleus

Division Phaeophyta
Brown algae

All marine, except three freshwater species.
Nearly all the members of this division occur in colder waters, predominating in the Arctic and Antarctic oceans. They are also very common round the coasts of Britain. Many species are found in the intertidal zone, where there is often a distinct vertical zonation.

All multicellular; simple or branched filamentous, or thalloid; mostly macroscopic.
The filamentous forms can be very small, consisting of a few cells only, whereas the sporophytes of many of the larger seaweeds can grow to several metres in length. The larger forms are clearly differentiated into holdfast, stipe and lamina.

Eucaryotic cells, with distinct cell walls, made up of an inner cellulose portion and an outer gelatinous portion containing alginic acid. The protoplasts are uninucleate, with prominent nuclei, and usually more than one chromatophore.
Most cells contain a number of small vacuoles and colourless, refractive bodies called fucosan vesicles. These vesicles were thought at one time to be linked with food storage, or to be waste products, but it is now thought more likely that they are concerned with the active metabolism of cells.

Flagella present in most genera, normally 2, of unequal length, one pantonematic, one acronematic, inserted laterally.
Biflagellate zoospores are present in all the orders except the Fucales. It has been shown that the longer flagellum is usually pantonematic and forward-pointing, whereas the shorter flagellum is acronematic and directed backwards. Biflagellate gametes are also produced.

Photosynthetic pigments include chlorophylls a and c, beta-carotene and a number of xanthophylls, notably fucoxanthin. There are no bilo-proteins present.

Chromatophores are usually numerous, discoid and yellow-brown, without any pyrenoids.
There appears to be only one order, the Ectocarpales, in which there is any great variety in chromatophore structure. The chromatophores may be plate-like and single, ribbon-shaped or discoid.

Storage products are the carbohydrates laminarin and mannitol, some fat, but no starch.

Vegetative multiplication can occur by the fragmentation of the attached thallus at the juvenile or adult stage, or by detachment of fragments of thallus which float away.

In the former case, a single attached thallus may fragment into 2 or more portions which remain attached, thus forming a cluster of individuals. A good example of the latter case is seen in the free-floating species of *Sargassum*, which is abundant in the Gulf Stream and the Sargasso Sea; no other form of reproduction has been observed for many of these species.

Asexual reproduction occurs by means of zoospores or aplanospores.

Many members of the division have a well-defined alternation of generations and the sporophyte generation is always diploid. Typically, zoosporangia occur on the sporophyte plants and are of two types: unilocular and plurilocular. In the plurilocular sporangia, the zoospores are formed by mitosis and are diploid, germinating to produce diploid sporophyte plants. Meiosis occurs during the formation of the zoospores in the unilocular sporangia, with the result that they are haploid and germinate to give haploid gametophyte plants.

Sexual reproduction is isogamous, anisogamous or oogamous.

Many members of this division produce multicellular gametangia which resemble plurilocular sporangia and, where there is an isomorphic alternation of generations, it is difficult to decide whether a structure contains zoospores or gametes. Isogamy, or anisogamy, depending on the species, occurs between motile gametes produced from multicellular gametangia. Where oogamy occurs, there is fusion of a small, biflagellate antherozoid with a large, non-flagellate oosphere.

In the brown algae, there is only one class, the Phaeophyceae, divided into three sub-classes.

Sub-class Isogeneratae: shows isomorphic alternation of generations, where gametophyte and sporophyte are alike in vegetative structure. There are 5 orders in this sub-class.

Example: Order Ectocarpales   *Ectocarpus*

Sub-class Heterogeneratae: the alternation of generations is heteromorphic, the sporophyte stage usually consisting of a complex thallus, which is either formed by aggregations of filaments (pseudo-parenchymatous) or by longitudinal and transverse divisions of cells (parenchymatous).

Gametophyte stages consist of a simple, filamentous thallus. There are 5 orders in this sub-class.

    Example: Order Laminariales   *Laminaria*

Sub-class Cyclosporae: no alternation of generations. The gametophyte generation is represented only by the gametes which are formed on the thallus of the sporophyte generation. There is only one order in this sub-class.

    Example: Order Fucales   *Fucus*
                                            Ascophyllum

Order Ectocarpales

Marine, with world-wide distribution; epiphytic on larger brown seaweeds such as *Ascophyllum* and *Fucus* species.

Branched, filamentous thallus, often heterotrichous with a well-marked prostrate and erect system. Sometimes the branches aggregate to form a pseudo-parenchymatous thallus.

Chromatophores are either few and band-shaped, or many and discoid; cells are uninucleate.

Asexual reproduction is by zoospores produced in unilocular or plurilocular sporangia. Unilocular sporangia are developed from the terminal cells of short lateral branches of the filament. The single nucleus divides meiotically, and then the daughter nuclei divide repeatedly until there are 32 to 64 nuclei present in the sporangium. After nuclear division is completed, the protoplasm divides up to form pyriform, biflagellate zoospores, each one haploid. Plurilocular sporangia develop in a similar position on the filament, but vertical and transverse divisions occur. This results in the formation of a multicellular structure, containing several hundred cells. Each cell produces a biflagellate zoospore, which is diploid as all the divisions have been mitotic. Haploid zoospores develop into gametophytes, bearing multicellular gametangia, whereas diploid zoospores develop into sporophytes, which can bear either unilocular or plurilocular sporangia.

Sexual reproduction is usually by isogamous fusion of gametes from the multicellular gametangia. In some genera, fusion is anisogamous. Sometimes gametes may develop parthenogenetically into gametophyte plants if fusion fails to occur.

There is an isomorphic alternation of generations in this order. The haploid gametophyte and diploid sporophyte stages are identical, except that the gametophyte bears only multicellular gametangia, while the sporophyte bears both unilocular and plurilocular sporangia.

Ectocarpus

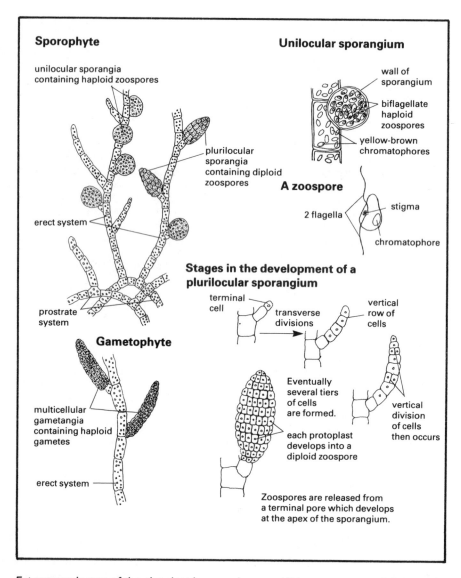

Sporophyte

unilocular sporangia containing haploid zoospores

plurilocular sporangia containing diploid zoospores

erect system

prostrate system

Unilocular sporangium

wall of sporangium

biflagellate haploid zoospores

yellow-brown chromatophores

A zoospore

2 flagella

stigma

chromatophore

Gametophyte

multicellular gametangia containing haploid gametes

erect system

Stages in the development of a plurilocular sporangium

terminal cell

transverse divisions

vertical row of cells

Eventually several tiers of cells are formed.

each protoplast develops into a diploid zoospore

vertical division of cells then occurs

Zoospores are released from a terminal pore which develops at the apex of the sporangium.

Ectocarpus is one of the simplest brown algae and it is a common epiphyte on *Fucus* and *Ascophyllum* spp. round the coasts of Britain. It forms a thick growth on its host plant, often penetrating the tissues. It seems that only the sporophyte stages of *Ectocarpus* are found in Britain, whereas the gametophyte stage is abundant in the Mediterranean, so the alternation of generations in the life cycle is not a very rigid one.

Ectocarpus

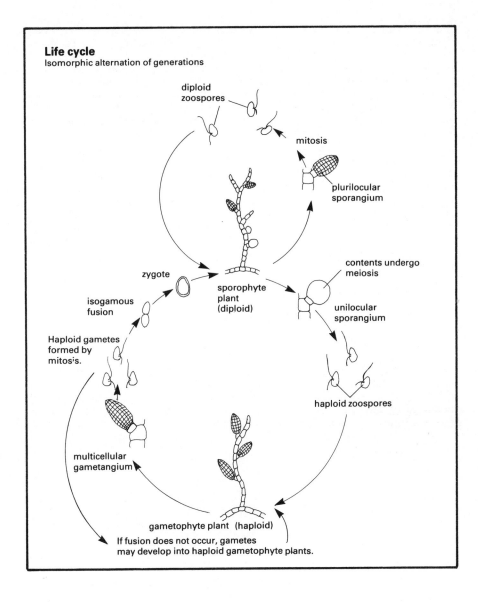

Life cycle
Isomorphic alternation of generations

diploid zoospores

mitosis

plurilocular sporangium

zygote

sporophyte plant (diploid)

contents undergo meiosis

isogamous fusion

unilocular sporangium

Haploid gametes formed by mitosis.

haploid zoospores

multicellular gametangium

gametophyte plant (haploid)

If fusion does not occur, gametes may develop into haploid gametophyte plants.

Order Laminariales

Marine, in the colder waters; do not occur in tropical or warm temperate seas. They are found in sub-littoral zones and are only rarely exposed at low tide.

There is regular alternation of a large, complex, parenchymatous sporophyte generation with a small, sparsely branched, filamentous gametophyte generation. The sporophyte thallus consists of a holdfast, stipe and lamina, and some of the largest species may be up to 30 metres long, e.g. *Macrocystis pyrifera*. Growth is initiated by a meristem situated between the stipe and the lamina.

There is a high degree of differentiation of the tissues of the sporophyte into superficial cells (epidermis), cortex and medulla. The medulla is made up of vertical, elongated, unbranched filaments, many of which have inflated ends, called 'trumpet hyphae'. There is evidence that these filaments have a conducting function and may be comparable to the sieve cells of higher plants. The epidermis may be one or two cells thick; the cells are small, cuboidal and contain many discoid chromatophores. Mucilage ducts are found in the cortex of many species. These canals are organized into a branched system, and each one is lined with mucilage-secreting cells.

On the sporophyte, sporangia are formed in sori on the surface of the fronds. Sporangia are of the unilocular type only, so that all the zoospores are haploid and will germinate to give rise to gametophyte plants.

The male gametophytes are multicellular and antheridia develop at the tips of short lateral branches. Inside each antheridium, one biflagellate antherozoid develops. The cells of the female gametophytes are bigger than those of the males, and oogonia may develop from all the cells. A single egg is produced in each oogonium. Male gametes are released and swim to the partially extruded eggs. Fertilization occurs and the diploid zygote begins its development into a new sporophyte plant whilst still attached to the female gametophyte.

Laminaria spp.

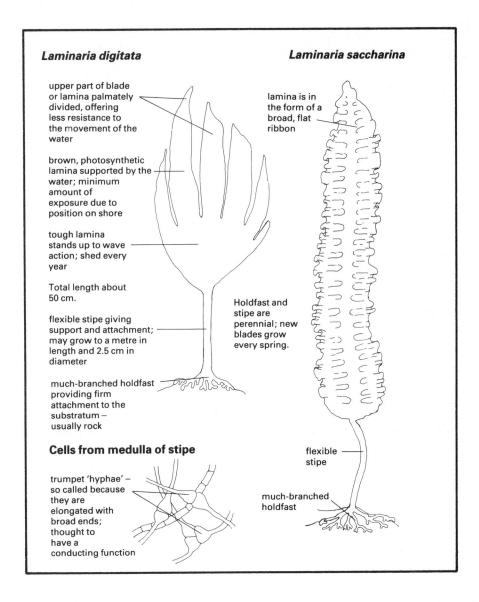

Laminaria digitata

Laminaria saccharina

upper part of blade or lamina palmately divided, offering less resistance to the movement of the water

lamina is in the form of a broad, flat ribbon

brown, photosynthetic lamina supported by the water; minimum amount of exposure due to position on shore

tough lamina stands up to wave action; shed every year

Total length about 50 cm.

flexible stipe giving support and attachment; may grow to a metre in length and 2.5 cm in diameter

Holdfast and stipe are perennial; new blades grow every spring.

much-branched holdfast providing firm attachment to the substratum – usually rock

Cells from medulla of stipe

flexible stipe

trumpet 'hyphae' – so called because they are elongated with broad ends; thought to have a conducting function

much-branched holdfast

55

Laminaria

DIVISION PHAEOPHYTA

CLASS PHAEOPHYCEAE

ORDER LAMINARIALES

GENUS *Laminaria*

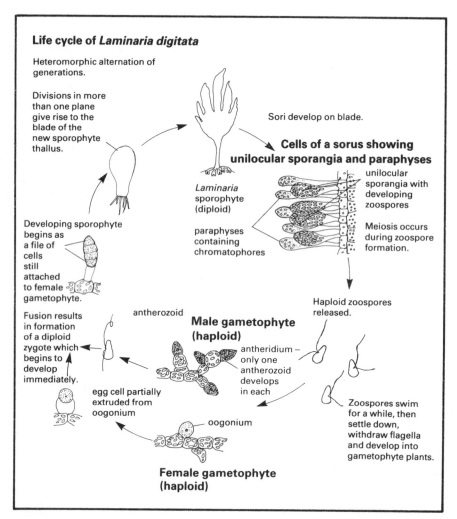

Life cycle of *Laminaria digitata*

Heteromorphic alternation of generations.

Divisions in more than one plane give rise to the blade of the new sporophyte thallus.

Sori develop on blade.

Cells of a sorus showing unilocular sporangia and paraphyses

Laminaria sporophyte (diploid)

unilocular sporangia with developing zoospores

paraphyses containing chromatophores

Meiosis occurs during zoospore formation.

Developing sporophyte begins as a file of cells still attached to female gametophyte.

Haploid zoospores released.

antherozoid

Male gametophyte (haploid)

Fusion results in formation of a diploid zygote which begins to develop immediately.

antheridium – only one antherozoid develops in each

egg cell partially extruded from oogonium

oogonium

Zoospores swim for a while, then settle down, withdraw flagella and develop into gametophyte plants.

Female gametophyte (haploid)

On rocky shores, *Laminaria* species are found in the lowest zones, uncovered only by the spring tides. They are the largest brown seaweeds found round the coasts of Britain.

The Laminariales, or kelps, provide the raw materials for the Japanese dish, kombu. The plants are harvested, dried and then cut up in a variety of ways before being cooked and eaten. In the past, the kelps have provided sources of soda, potash and iodine. Today, their importance is in the production of alginic acid. The soluble salts of alginic acid are non-toxic, viscous and readily form gels. They are used in the manufacture of ice-cream, packet soups, cosmetics, surgical dressings and emulsion paints.

Order Fucales

Members of this order are restricted to cool waters in both the northern and the southern hemispheres; none are found in tropical seas. Most genera grow attached to rocks in the intertidal zone, often showing very definite vertical zonation. In some species (*Sargassum*), the thallus is free-floating.

It would seem that, unlike many of the related Phaeophyceae, there is no obvious alternation of generations in the Fucales. The diploid plants undergo meiosis and the products behave like gametes, fusing to give a diploid zygote which develops into a diploid thallus. This type of life cycle is thought to have been derived from an alternation of generations in which the haploid gametophyte generation has been lost.

Growth of the thallus is by an apical cell situated at the tip of each dichotomy. The thailus is differentiated into an outer layer of close-fitting photosynthesizing cells, a middle cortex and an inner medulla of long, conducting filaments embedded in mucilage.

Some regeneration of broken pieces of thallus occurs, but there is no spore production or other means of asexual reproduction.

Sexual reproduction is oogamous and the reproductive organs develop at the tips of the fronds in special cavities called conceptacles. In some species, both antheridia and oogonia are formed on the same plant, while in others they are borne on separate plants. Inside the antheridia, 64 biflagellate antherozoids develop, differing from the motile cells in other Phaeophyceae in that the posterior flagellum is longer than the anterior one. The number of oospheres which develop in an oogonium varies according to the genus: 8 in *Fucus*, 4 in *Ascophyllum*, 2 in *Pelvetia* and 1 in *Sargassum*. After fertilization, the diploid zygote will begin development into a new thallus without a period of dormancy.

Fucus spp.

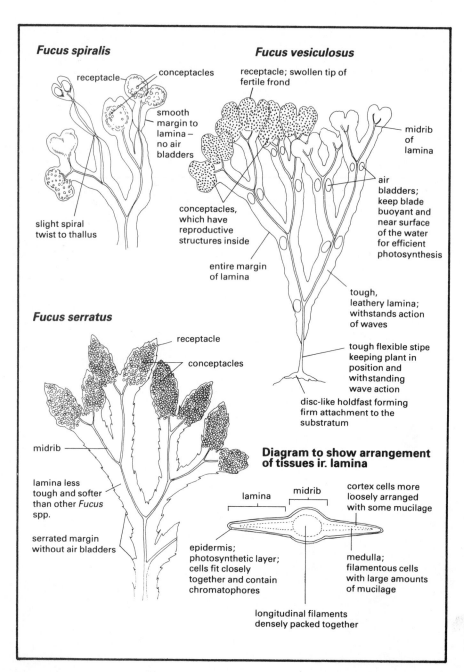

Fucus spiralis

receptacle

conceptacles

smooth margin to lamina – no air bladders

slight spiral twist to thallus

Fucus vesiculosus

receptacle; swollen tip of fertile frond

midrib of lamina

air bladders; keep blade buoyant and near surface of the water for efficient photosynthesis

conceptacles, which have reproductive structures inside

entire margin of lamina

tough, leathery lamina; withstands action of waves

tough flexible stipe keeping plant in position and withstanding wave action

disc-like holdfast forming firm attachment to the substratum

Fucus serratus

receptacle

conceptacles

midrib

lamina less tough and softer than other *Fucus* spp.

serrated margin without air bladders

Diagram to show arrangement of tissues in lamina

lamina

midrib

cortex cells more loosely arranged with some mucilage

epidermis; photosynthetic layer; cells fit closely together and contain chromatophores

medulla; filamentous cells with large amounts of mucilage

longitudinal filaments densely packed together

DIVISION PHAEOPHYTA

CLASS PHAEOPHYCEAE

ORDER FUCALES

GENUS *Fucus*

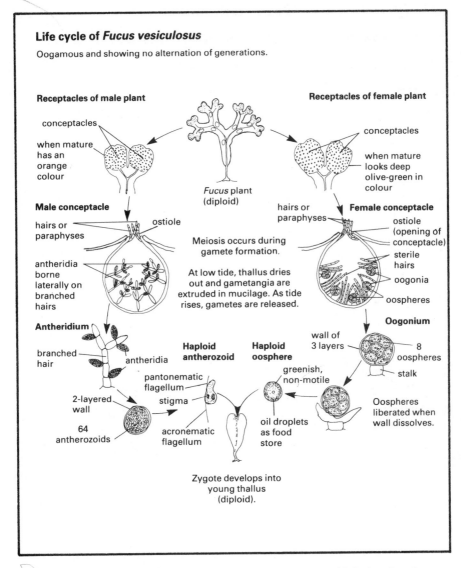

Life cycle of *Fucus vesiculosus*

Oogamous and showing no alternation of generations.

Receptacles of male plant

conceptacles

when mature
has an
orange
colour

Fucus plant
(diploid)

Receptacles of female plant

conceptacles

when mature
looks deep
olive-green in
colour

Male conceptacle

hairs or
paraphyses

ostiole

antheridia
borne
laterally on
branched
hairs

Meiosis occurs during
gamete formation.

At low tide, thallus dries
out and gametangia are
extruded in mucilage. As tide
rises, gametes are released.

hairs or
paraphyses

Female conceptacle

ostiole
(opening of
conceptacle)

sterile
hairs

oogonia

oospheres

Antheridium

branched
hair

antheridia

2-layered
wall

64
antherozoids

**Haploid
antherozoid**

pantonematic
flagellum

stigma

acronematic
flagellum

**Haploid
oosphere**

greenish,
non-motile

oil droplets
as food
store

Oogonium

wall of
3 layers

8
oospheres

stalk

Oospheres
liberated when
wall dissolves.

Zygote develops into
young thallus
(diploid).

Three *Fucus* species are fairly common round the shores of Britain, showing a distinct zonation in the littoral zone. *Fucus spiralis*, the spiral wrack, grows high up on the shore, enduring most exposure as it is only submerged by the spring tides. *Fucus vesiculosus*, the bladder wrack, is found in the middle zone of the shore and *Fucus serratus*, the serrated wrack, grows lower down near the low tide level, just above the *Laminaria* zone.

Some common brown algae found round the coasts of Britain

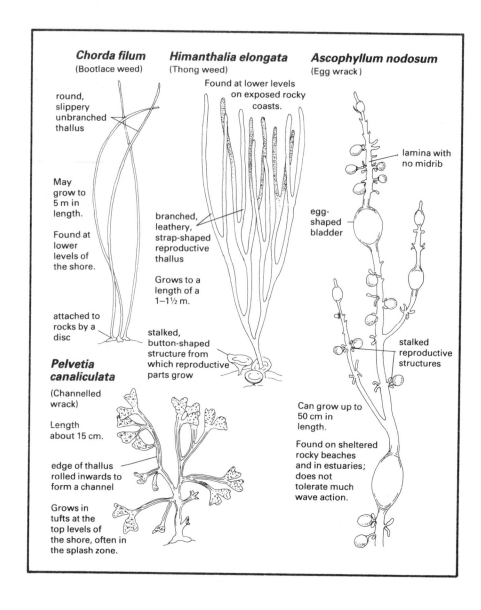

Chorda filum
(Bootlace weed)

round, slippery unbranched thallus

May grow to 5 m in length.

Found at lower levels of the shore.

attached to rocks by a disc

Pelvetia canaliculata

(Channelled wrack)

Length about 15 cm.

edge of thallus rolled inwards to form a channel

Grows in tufts at the top levels of the shore, often in the splash zone.

Himanthalia elongata
(Thong weed)

Found at lower levels on exposed rocky coasts.

branched, leathery, strap-shaped reproductive thallus

Grows to a length of a 1–1½ m.

stalked, button-shaped structure from which reproductive parts grow

Ascophyllum nodosum
(Egg wrack)

lamina with no midrib

egg-shaped bladder

stalked reproductive structures

Can grow up to 50 cm in length.

Found on sheltered rocky beaches and in estuaries; does not tolerate much wave action.

Division Rhodophyta
Red algae

Mostly marine, but a few species are freshwater or terrestrial.
> The freshwater species are restricted to cold, fast-flowing water in waterfalls, rapids and mill dams. The marine species are ubiquitous, though relatively few are found in polar seas, and there is a definite pattern of distribution of species relative to temperature, together with a vertical zonation in the intertidal and sub-tidal zones. The latter distribution seems to be linked to light intensity. The majority of the marine species are attached to rocks or other substrata, and a significant number grow epiphytically on other marine algae.

Mostly filamentous or thalloid, rarely unicellular.
> The structure of the thallus is basically filamentous; filaments may be simple or branched, free or compacted together to form a pseudo-parenchymatous plant body.

Eucaryotic cells; cell walls contain cellulose and pectic compounds.
> Most cells are uninucleate with a large central vacuole. Some species have pits, which appear to allow cytoplasmic communication between adjacent cells.

Flagella absent.

Photosynthetic pigments include chlorophylls *a* and *d*, carotenes, one xanthophyll and the bilo-proteins R-phycoerythrin and R-phycocyanin.
> The principal pigments are chlorophyll *a*, beta-carotene and R-phycoerythrin; the latter masking the other pigments and giving the plants their characteristic red colour. Many members of this division are found at depths of water where blue light predominates and the phycoerythrin enables more efficient use of available light for photosynthesis.

Chromatophores vary from single, central and stellate in the more primitive genera, to many and discoid in the more advanced genera.
> In the simpler forms, there are dense structures of protein, which have been called pyrenoids, but they differ from the pyrenoids found in other algae in that they do not have starch grains around them. The more complex members of the division do not have these pyrenoids.

Storage product is Floridean starch.
> Floridean starch is a glucose polymer, differing from ordinary starch by containing more amylopectin. It appears to resemble glycogen, and it occurs as grains scattered in the cytoplasm.

Asexual reproduction may occur by monospores or neutral spores, according to the genus.

Sexual reproduction is by fusion of male cells (called spermatia) with the female organs (the carpogonia) followed by asexual spore formation. Alternation of generations occurs and the life cycles can be extremely complex.

The division contains one class, the Rhodophyceae, which is divided into 2 sub-classes.

Sub-class Bangioideae containing 15 genera and some 70 species.

Sub-class Floridae containing 375 genera and about 2500 species.

Sub-class Bangioideae	Sub-class Floridae
Unicellular, simple filaments, branched filaments, solid cylinders or expanded sheets.	Branched, filamentous thallus. In some genera, branches are free; in others they are intermingled in a gelatinous matrix or closely bound together to form a pseudoparenchymatous structure.
Very few pits or connections between cells.	Pits and cytoplasmic connections between adjacent cells are common.
Chromatophores commonly single, central and stellate; rarely, numerous, parietal and discoid.	Chromatophores usually numerous, parietal and discoid; rarely, single, central and stellate.
Growth of thallus is by intercalary cell division.	Growth of thallus is by apical cell division of the filaments. A single filament may give off laterals on all sides (monoaxial organization), or there may be a central core of axial filaments each giving off laterals (multiaxial).
Asexual reproduction by monospores formed singly in monosporangia, or by neutral spores developed directly from vegetative cells.	Asexual reproduction by monospores as in the Bangioideae.
In sexual reproduction, after fertilization, haploid carpospores are formed directly from the zygote, and germinate to give rise to the gametophyte plant.	In sexual reproduction, a carposporophyte is formed and the carpospores produced in some genera are haploid, germinating to produce the gametophyte. In most genera, carpospores are diploid and develop into tetrasporophytes, where meiosis takes place during the production of haploid tetraspores. On germination, a tetraspore will grow into a gametophyte plant.
Example: *Porphyra umbilicalis*	Examples: *Corallina officinalis* *Chondrus crispus* *Polysiphonia*

Porphyra umbilicalis

DIVISION RHODOPHYTA

CLASS RHODOPHYCEAE

SUBCLASS BANGIOIDEAE

GENUS *Porphyra*

Life cycle of *Porphyra umbilicalis*

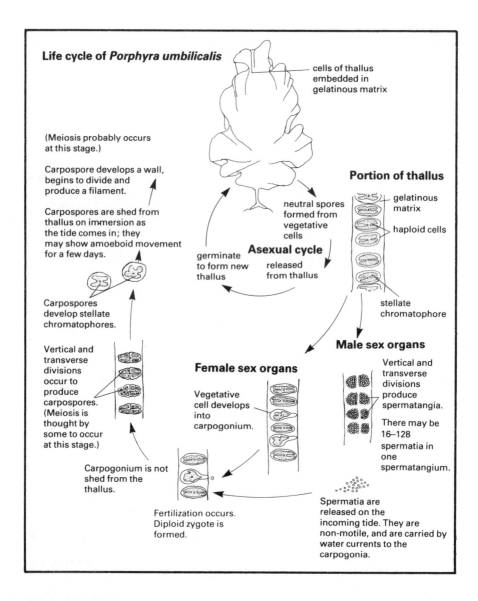

cells of thallus embedded in gelatinous matrix

(Meiosis probably occurs at this stage.)

Carpospore develops a wall, begins to divide and produce a filament.

Carpospores are shed from thallus on immersion as the tide comes in; they may show amoeboid movement for a few days.

Carpospores develop stellate chromatophores.

Vertical and transverse divisions occur to produce carpospores. (Meiosis is thought by some to occur at this stage.)

Carpogonium is not shed from the thallus.

Portion of thallus

gelatinous matrix

haploid cells

neutral spores formed from vegetative cells

Asexual cycle

germinate to form new thallus

released from thallus

stellate chromatophore

Male sex organs

Vertical and transverse divisions produce spermatangia.

There may be 16–128 spermatia in one spermatangium.

Female sex organs

Vegetative cell develops into carpogonium.

Fertilization occurs. Diploid zygote is formed.

Spermatia are released on the incoming tide. They are non-motile, and are carried by water currents to the carpogonia.

Porphyra umbilicalis grows attached to rocks high up on the shore, often amongst *Pelvetia*. It has a reddish-purple sheet-like thallus, one or two cells thick, and resembles *Ulva*.

Porphyra is used to make the purple laver 'bread', which is still popular in South Wales. A similar dish is eaten in China and Japan.

Some members of the Florideae

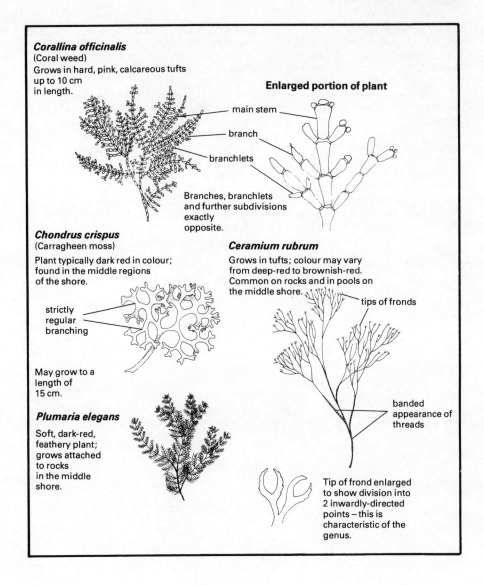

Corallina officinalis
(Coral weed)
Grows in hard, pink, calcareous tufts
up to 10 cm
in length.

Enlarged portion of plant

main stem

branch

branchlets

Branches, branchlets
and further subdivisions
exactly
opposite.

Chondrus crispus
(Carragheen moss)

Plant typically dark red in colour;
found in the middle regions
of the shore.

strictly
regular
branching

May grow to a
length of
15 cm.

Plumaria elegans

Soft, dark-red,
feathery plant;
grows attached
to rocks
in the middle
shore.

Ceramium rubrum

Grows in tufts; colour may vary
from deep-red to brownish-red.
Common on rocks and in pools on
the middle shore.

tips of fronds

banded
appearance of
threads

Tip of frond enlarged
to show division into
2 inwardly-directed
points – this is
characteristic of the
genus.

Fungi

There are more than 40 000 different species of fungi, occupying a wide range of habitats and exploiting a great variety of food sources. Many species are of considerable economic importance to man, causing spoilage of crops and stored foods, diseases, bringing about fermentation and acting as sources of antibiotics.

The fungi are mostly terrestrial organisms lacking chlorophyll and other photosynthetic pigments. The classification is based on the nature of the plant body; the forms which consist of naked masses of protoplasm, or plasmodia, being placed in the division Myxomycophyta (slime moulds), and the unicellular and filamentous forms being placed in the division Eumycophyta (true fungi). Further division of the latter group depends on the nature of the filaments, or hyphae, which make up the plant body, or mycelium.

It is proposed to deal in detail with representatives of the division Eumycophyta, and to give an outline only of the division Myxomycophyta.

It is worth recording that the term 'fungus', in common with the term 'alga', is no longer considered to have any taxonomic significance, but remains a convenient descriptive word when referring to members of the Eumycophyta.

Division Myxomycophyta
Slime moulds

Terrestrial; parasitic or saprophytic.

Most slime moulds occur on dead leaves, decaying logs and other damp organic matter. Some species will develop on bark, if it is kept moist, and there are species present in most soil. They are especially common in late spring and early autumn after heavy rain. Two or three genera are of economic importance as parasites of crops, but the majority are saprophytes, or feed on bacteria and other micro-organisms which are engulfed by the protoplasm.

Plant body a plasmodium or pseudoplasmodium.

A plasmodium is a single, large, multinucleate, naked mass of protoplasm; a pseudoplasmodium is the result of the aggregation of a large number of small, naked, uninucleate protoplasts. In the latter case, each uninucleate protoplast retains its individuality, no fusion occurs and the whole structure will break up into its constituent units when placed in water.

Reproduction by small, uninucleate spores.

In most genera, the spores are developed either in or on a fructification, or fruiting body, which develops from the plasmodium. Large numbers of spores are produced and germinate to give rise either to biflagellate swarm cells, or non-flagellate amoeboid cells. In some genera, the swarm cells function as gametes and fuse in pairs. The resulting diploid zygote will divide mitotically to give rise to a new vegetative plasmodium. In other genera, the swarm cells develop into plasmodia. In this case, each plasmodium is haploid, and at a later stage more swarm cells will be produced and fuse in pairs, forming a diploid zygote. Meiosis is thought to occur just prior to spore formation.

The Myxomycophyta show characteristics of both plants and animals, and there has been considerable debate about their inclusion in the plant kingdom. Despite their resemblance to giant Amoebae and the absence of cell walls in the vegetative body, their reproductive structures are plant-like. On these grounds, the organisms are classified as plants, and are thought to be representative of a primitive type of fungal organization.

The division Myxomycophyta is made up of three classes.

Class Myxomycetae: called the true slime moulds. The members of this class have a true plasmodium which behaves as a unit, and which may be up to several centimetres in diameter.

Class Acrasieae: pseudoplasmodia are formed and no flagellated swarm cells occur. Many genera are found on dung and have been isolated from soils with a high organic content.

Class Plasmodiophoreae: have a plasmodial type of thallus, but no fruiting structures are formed and the resting spores are formed directly from the plasmodium. They are parasitic on higher plants, fungi and algae.

Example: *Plasmodiophora brassicae* which causes clubroot of the Cruciferae.

Plasmodiophora brassicae

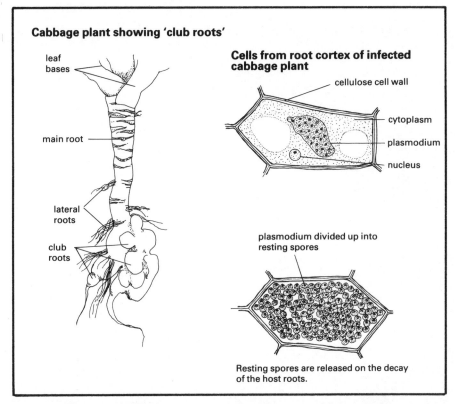

Cabbage plant showing 'club roots'

leaf bases

main root

lateral roots

club roots

Cells from root cortex of infected cabbage plant

cellulose cell wall

cytoplasm

plasmodium

nucleus

plasmodium divided up into resting spores

Resting spores are released on the decay of the host roots.

Plasmodiophora brassicae is a parasite and causes a disease known as club root in members of the Cruciferae, especially cabbages and related plants. The disease causes characteristic club-like swellings of infected roots. These swellings, when cut open, show a mottled appearance, and the aerial parts of infected plants become yellow and show limited growth due to the interference with water and salt uptake.

Infection occurs via the young root hairs, and the fungus spreads into the cortical cells, which are stimulated to divide. This increase in the number of cortical cells forms the characteristic swellings and if the cambial cells become infected the formation of club roots is especially rapid. Resting spores are produced and released into the soil on the decay of the host roots. These spores have a smooth cell wall of chitin, and can remain viable for as long as 7 or 8 years.

The control of the disease is difficult because of the longevity of the spores, making crop rotation of little value. Control can be achieved by liming the soil, growing disease-resistant varieties of cultivated plants or the use of fungicides. Mercury, copper compounds, formalin and chlorinated nitrobenzenes have been used as seed treatments with good results.

Plasmodiophora brassicae

DIVISION MYXOMYCOPHYTA
CLASS PLASMODIOPHOREAE
GENUS *Plasmodiophora*

Life cycle of *Plasmodiophora brassicae*

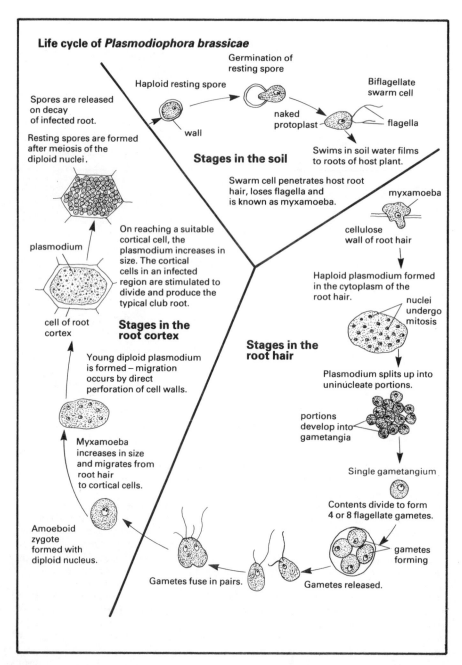

Germination of resting spore

Haploid resting spore

Biflagellate swarm cell

Spores are released on decay of infected root.

naked protoplast

flagella

Resting spores are formed after meiosis of the diploid nuclei.

wall

Stages in the soil

Swims in soil water films to roots of host plant.

Swarm cell penetrates host root hair, loses flagella and is known as myxamoeba.

myxamoeba

plasmodium

On reaching a suitable cortical cell, the plasmodium increases in size. The cortical cells in an infected region are stimulated to divide and produce the typical club root.

cellulose wall of root hair

Haploid plasmodium formed in the cytoplasm of the root hair.

nuclei undergo mitosis

cell of root cortex

Stages in the root cortex

Stages in the root hair

Young diploid plasmodium is formed – migration occurs by direct perforation of cell walls.

Plasmodium splits up into uninucleate portions.

portions develop into gametangia

Myxamoeba increases in size and migrates from root hair to cortical cells.

Single gametangium

Contents divide to form 4 or 8 flagellate gametes.

Amoeboid zygote formed with diploid nucleus.

gametes forming

Gametes fuse in pairs.

Gametes released.

Division Eumycophyta
True fungi

Terrestrial or aquatic; parasitic or saprophytic.
> The great majority of the saprophytic Eumycophyta are terrestrial, growing in the soil or on the remains of animals or plants, and only a few saprophytic species are aquatic. Many species are parasitic and their hosts range from simple algae to angiosperms. Some species form symbiotic relationships with the higher plants, called mycorrhizae, and other species are the mycobionts of lichens.

Plant body unicellular or filamentous.
> Most Eumycophyta have a branching, filamentous type of thallus called a mycelium. The individual filaments are known as hyphae and the mycelium may be coenocytic, with no transverse septa, or consist of uninucleate, binucleate or multinucleate cells. The hyphae may form a loose network or weft, or they may be compacted together to form a pseudoparenchymatous structure. The cell walls often contain chitin.

Heterotrophic; no photosynthetic pigments present.
> Saprophytic species grow on their food substrate, secreting enzymes from the mycelium, and digestion is extra-cellular. The soluble products of digestion diffuse into the hyphae. The parasitic species obtain their food from the host organism. Facultative parasites can live both on the host and on organic remains, but the obligate parasites are restricted to living hosts only. These species have special absorptive hyphae, or haustoria, which penetrate into the cells of the host in order to obtain food.

Asexual reproduction by means of sporangiospores or conidiospores.
> Spore production is prolific, and in many groups it is the most general method of reproduction. Spores may be formed directly on the mycelium or in sporangia. Motile sporangiospores are common in aquatic genera. They are called zoospores and each possesses one or two flagella. Terrestrial genera produce non-motile sporangiospores, or aplanospores, which are dependent on air movements for their dispersal. Conidiospores are often borne in chains at the apex of a specialized hypha, the conidiophore, and are common in terrestrial species.

Sexual reproduction by fusion of gametes or gamete nuclei.
> The gametes may be produced in special sex organs, or gametangia, and sexual reproduction can be oogamous or by the conjugation of similar non-motile gametes. In some genera, a zygote is produced after fusion of two specialised hyphae, or by the fusion of two nuclei in a vegetative cell. Many members of this division exhibit heterothallism, where the gametes or gamete nuclei must come from two different mycelia of compatible mating strains. If the gametes or gamete nuclei are derived from the same mycelium, the fungus is said to be homothallic, or self-fertile. Many genera produce fruiting structures, or sporophores, of distinctive shape and size after sexual fusion.

The fossil record of the Eumycophyta is scanty, although some fossils resembling Phycomycetes have been found in the Devonian, and many more are known from later periods. However, the origins of the fungi are obscure and it is difficult to determine any evolutionary trends within the division.

The classification is based on the nature of the mycelium, whether it is coenocytic or septate, and the nature of the spore-producing structures. Four classes are recognized.

Class Phycomycetes: contains about 1400 species.

Class Ascomycetes: contains about 15 500 species.

Class Basidiomycetes: contains about 15 000 species.

Class Deuteromycetes (Fungi Imperfecti): these are the fungi for which no 'perfect' or sexual stage has yet been discovered. The number of species in this class is changing all the time, but there are about 11 000 at present.

Class Phycomycetes

Members of this class may be terrestrial or aquatic, and are mostly saprophytic. There are many aquatic forms, both parasites and saprophytes, in this class. Some parasitic genera live on algae and other aquatic Phycomycetes, whereas the aquatic saprophytes are found on vegetable remains and dead fish and insects. The terrestrial genera are common inhabitants of the soil, and the spores of many species are always present in the air. Many parasitic forms cause serious mildews and blights of crop plants, and the saprophytes can cause spoilage of food and rot fruit and vegetables in store.

The most primitive members of this class are unicellular, but the majority possess a coenocytic, multinucleate mycelium. Septa, or cross walls, are usually formed in connection with the development of reproductive structures. Cell walls are present and may be of cellulose or fungal chitin. Food reserves are usually oil, occurring as droplets in the cytoplasm.

Asexual reproduction is by sporangiospores or conidiospores. Sporangia are usually borne at the tips of special hyphae called sporangiophores. The contents of each sporangium divide to form an indefinite number of sporangiospores, which may become non-motile aplanospores, or develop one or two flagella to become motile zoospores. Zoospores are produced by aquatic species and also by some terrestrial species, in which case they are dependent on surface films of water for their dispersal. Sporangiospores germinate, when conditions are suitable, into new mycelia.

Sexual reproduction is by fusion of gametes or gamete nuclei, which are produced in unicellular gametangia, and fusion of gametes is soon followed by fusion of nuclei in this class. Sexual reproduction is often oogamous, though isogamy and anisogamy do occur in the more primitive genera. In the more advanced forms, the gametes are reduced to the undifferentiated contents of gametangia, and sexual reproduction is achieved by conjugation of two gametangia. The members of this class may be homothallic or heterothallic. The resulting diploid zygote usually develops into a thick-walled resting structure, called a zygospore in some groups. Following a period of dormancy the zygospore will germinate, in favourable conditions, to give rise to a new thallus, a sporangiophore bearing a sporangium, or directly to zoospores. Meiosis is thought to occur prior to the germination of the zygospore in most species.

Three sub-classes are recognized, and the classification is based on the presence or absence of motile stages in the life cycle.

Sub-class Uniflagellatae: most genera form zoospores and gametes which have a single, acronematic, posteriorly inserted flagellum.

Sub-class Biflagellatae: zoospores biflagellate; 1 acronematic, 1 pantonematic flagellum.
    Order Peronosporales   *Pythium debaryanum*
                                  Phytophthora infestans

Sub-class Aflagellatae: no flagellate stages in the life cycle.
    Order Mucorales   *Mucor* spp.
                                   Rhizopus nigricans

Order Peronosporales

The genera in this order are terrestrial or aquatic, and many are parasitic. There is a group of highly-specialized obligate parasites, which cause serious diseases to crop plants, and whose life cycles have been studied in detail.

The mycelium is coenocytic, with well-developed, stout, branching hyphae typical of the class. The parasitic species produce haustoria, which penetrate into the host's cells to absorb nutrients. The haustoria may be branched, knob-like or elongated.

The sporangia are typically oval or lemon-shaped and produce zoospores, which are kidney-shaped and biflagellate. In some cases, the whole sporangium acts as a spore and will germinate to produce a germ tube, instead of forming zoospores.

In sexual reproduction, the antheridia and oogonia are borne on the same or on different hyphae. The oogonia are usually globose, containing a single uninucleate or multinucleate oosphere. The antheridia are uninucleate or multinucleate. Contact between an antheridium and an oogonium is made and a fertilization tube is formed by the antheridium. This grows through the oogonial wall to the oosphere, and the male nucleus or nuclei pass through the tube. After fertilization, the oosphere develops a thick wall and becomes an oospore. After a period of dormancy, the oospore germinates by producing zoospores or by producing germ tubes, which later develop sporangia.

Pythium
debaryanum

DIVISION EUMYCOPHYTA

CLASS PHYCOMYCETES

SUB-CLASS BIFLAGELLATAE

ORDER PERONOSPORALES

GENUS *Pythium*

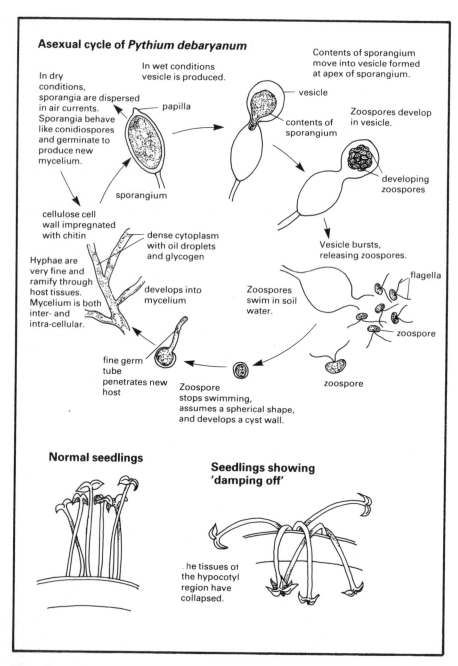

Asexual cycle of *Pythium debaryanum*

In dry conditions, sporangia are dispersed in air currents. Sporangia behave like conidiospores and germinate to produce new mycelium.

In wet conditions vesicle is produced.

Contents of sporangium move into vesicle formed at apex of sporangium.

vesicle

papilla

contents of sporangium

Zoospores develop in vesicle.

sporangium

developing zoospores

cellulose cell wall impregnated with chitin

Vesicle bursts, releasing zoospores.

dense cytoplasm with oil droplets and glycogen

Hyphae are very fine and ramify through host tissues. Mycelium is both inter- and intra-cellular.

develops into mycelium

Zoospores swim in soil water.

flagella

zoospore

fine germ tube penetrates new host

zoospore

Zoospore stops swimming, assumes a spherical shape, and develops a cyst wall.

Normal seedlings

Seedlings showing 'damping off'

. he tissues of the hypocotyl region have collapsed.

Pythium debaryanum

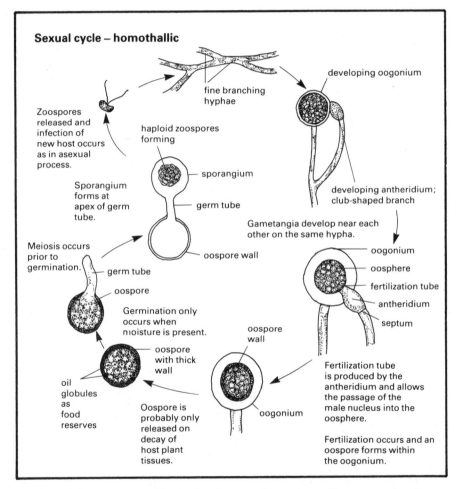

Sexual cycle – homothallic

developing oogonium

fine branching hyphae

Zoospores released and infection of new host occurs as in asexual process.

haploid zoospores forming

sporangium

Sporangium forms at apex of germ tube.

germ tube

developing antheridium; club-shaped branch

Gametangia develop near each other on the same hypha.

Meiosis occurs prior to germination.

oospore wall

germ tube

oospore

oogonium

oosphere

fertilization tube

antheridium

septum

Germination only occurs when moisture is present.

oospore wall

oospore with thick wall

oil globules as food reserves

Oospore is probably only released on decay of host plant tissues.

oogonium

Fertilization tube is produced by the antheridium and allows the passage of the male nucleus into the oosphere.

Fertilization occurs and an oospore forms within the oogonium.

Pythium debaryanum can live in the tissues of young seedlings as a parasite, causing a disease known as 'damping-off'. It causes the tissues in the hypocotyl region to collapse and the seedling falls over. After the seedling has been killed, *Pythium* can continue to live saprophytically on the dead tissues, and is therefore an example of a facultative parasite. *Pythium* spp. are common members of the soil microflora, feeding saprophytically on plant remains.

Infection of the seedlings can be from spores or mycelia in the soil, or on neighbouring plants. The mycelium can penetrate through cell walls or enter via stomata. Once inside the host, the mycelium grows in the intercellular spaces, sending branches into the cells. Asexual reproduction occurs soon after infection and the disease spreads rapidly, particularly in the humid conditions found in greenhouses.

Phytophthora infestans

DIVISION EUMYCOPHYTA

CLASS PHYCOMYCETES

SUB-CLASS BIFLAGELLATAE

ORDER PERONOSPORALES

GENUS *Phytophthora*

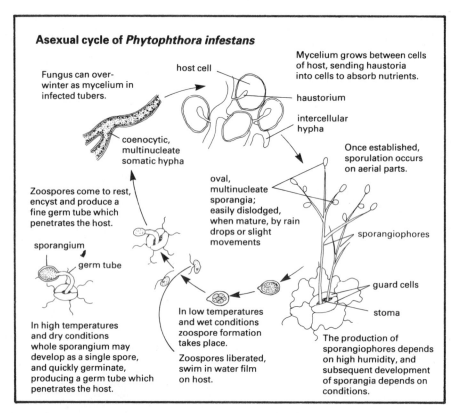

Asexual cycle of *Phytophthora infestans*

Fungus can over-winter as mycelium in infected tubers.

host cell

Mycelium grows between cells of host, sending haustoria into cells to absorb nutrients.

haustorium

intercellular hypha

coenocytic, multinucleate somatic hypha

Once established, sporulation occurs on aerial parts.

Zoospores come to rest, encyst and produce a fine germ tube which penetrates the host.

oval, multinucleate sporangia; easily dislodged, when mature, by rain drops or slight movements

sporangiophores

sporangium

germ tube

guard cells

stoma

In high temperatures and dry conditions whole sporangium may develop as a single spore, and quickly germinate, producing a germ tube which penetrates the host.

In low temperatures and wet conditions zoospore formation takes place.

Zoospores liberated, swim in water film on host.

The production of sporangiophores depends on high humidity, and subsequent development of sporangia depends on conditions.

The genus *Phytophthora* includes a large number of species which parasitise specific hosts causing soft rots, and hence is of considerable economic importance. *Phytophthora infestans* was responsible for the Irish potato famine, due to the failure of the entire potato crop two years running. This famine caused thousands of deaths and widespread emigration from Ireland to the United States of America in the middle of the nineteenth century.

Potato blight still occurs in Britain, particularly in the north and west where suitable conditions of humidity may prevail during the growing season. The asexual cycle is a very rapid means of reproduction, as only a few days are needed for the fungus to become established in the aerial parts of the plant before it will produce sporangiophores and sporangia.

First signs of infection: small brown patches on the leaves; it is often possible to see downy white patches on the underside of the leaves. The patches spread rapidly in wet conditions.

If attack is heavy: whole plant may become brown and rotten due to secondary infection of diseased tissues by bacteria.

Infection in the tubers: this shows as discoloured patches, which are rusty brown and are under the skin.

Phytophthora infestans

Sexual cycle of *Phytophthora infestans*

Sexual reproduction occurs rarely under natural conditions and has mostly been studied in culture.

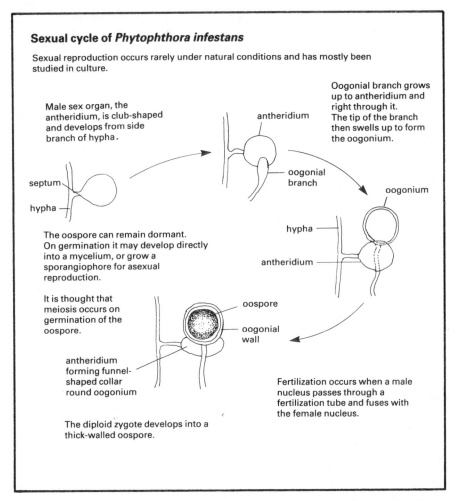

Male sex organ, the antheridium, is club-shaped and develops from side branch of hypha.

Oogonial branch grows up to antheridium and right through it. The tip of the branch then swells up to form the oogonium.

antheridium

oogonial branch

oogonium

septum

hypha

hypha

antheridium

The oospore can remain dormant. On germination it may develop directly into a mycelium, or grow a sporangiophore for asexual reproduction.

It is thought that meiosis occurs on germination of the oospore.

oospore

oogonial wall

antheridium forming funnel-shaped collar round oogonium

Fertilization occurs when a male nucleus passes through a fertilization tube and fuses with the female nucleus.

The diploid zygote develops into a thick-walled oospore.

The chief source of epidemics of potato blight is the planting of infected tubers, and it is also known that the fungus can survive in the soil for about a year.

In order to control the disease, infected plants should be burnt. Stored tubers should be kept dry, as this is less likely to cause spread. The parasite can be controlled by spraying potato plants with Bordeaux mixture at the correct times, with regard to the weather forecast; the spread of the disease being most rapid in wet conditions.

Planting of *Phytophthora*-free tubers is essential for the control of the disease, and the possibility of breeding disease-resistant varieties is obviously a solution to the problem. Unfortunately, *Phytophthora infestans* can exist in different forms and this can complicate the work of the plant breeders.

Order Mucorales

The genera in this order are mostly terrestrial saprophytes, living on a wide variety of organic substrates including dung, bread, cooked food, decaying plant and animal matter. There are a few parasitic genera which cause soft rots and some species cause diseases of crops in storage. Members of the Mucorales are common in soil and in the air.

The mycelium consists of stout, freely-branching coenocytic hyphae, which often become vacuolated, accumulating brown pigment and becoming septate when older. The hyphal walls are composed of fungal chitin.

Asexual reproduction is by means of aplanospores produced in sporangia, borne on sporangiophores, which may be simple or branched. The number of aplanospores within a sporangium is indefinite, and they may be uninucleate or multinucleate. The sporangium is cut off from the sporangiophore by a septum which may be flat or bulged. If the septum bulges, the sporangiophore has an inflated head, the columella, which extends into the sporangium.

Sexual reproduction is achieved by the fusion of two multinucleate gametangia, usually of equal size. Homothallic and heterothallic species occur within the order. The resulting zygote develops a thick wall and becomes a resting zygospore. There are no motile stages in this group.

Mucor spp.
Pin mould

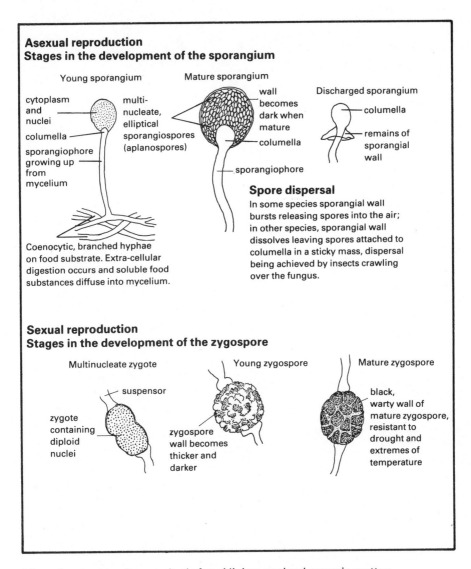

Asexual reproduction
Stages in the development of the sporangium

Young sporangium

cytoplasm and nuclei

columella

sporangiophore growing up from mycelium

Coenocytic, branched hyphae on food substrate. Extra-cellular digestion occurs and soluble food substances diffuse into mycelium.

multi-nucleate, elliptical sporangiospores (aplanospores)

Mature sporangium

wall becomes dark when mature

columella

sporangiophore

Discharged sporangium

columella

remains of sporangial wall

Spore dispersal
In some species sporangial wall bursts releasing spores into the air; in other species, sporangial wall dissolves leaving spores attached to columella in a sticky mass, dispersal being achieved by insects crawling over the fungus.

Sexual reproduction
Stages in the development of the zygospore

Multinucleate zygote

suspensor

zygote containing diploid nuclei

Young zygospore

zygospore wall becomes thicker and darker

Mature zygospore

black, warty wall of mature zygospore, resistant to drought and extremes of temperature

Mucor is a genus of saprophytic fungi living on dead organic matter, particularly the dung of horses and cattle, and spoiling human foods. It is possible to obtain examples of *Mucor* spp. by exposing a piece of moist brown bread to the atmosphere for 24 hours. If the bread is then covered, a fluffy white growth of mycelium appears after a few days, followed by the formation of a large number of tiny, black sporangia. A few species can be used commercially in the fermentation of sugar to alcohol.

Mucor spp.

Life cycle of *Mucor* spp. Asexual cycle

Occurs frequently when conditions for growth are favourable and results in the production of masses of spores which are available for rapid colonization of new food sources.

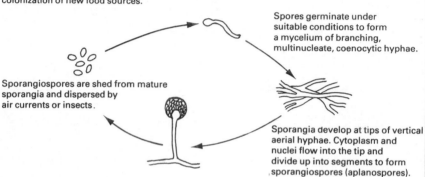

Spores germinate under suitable conditions to form a mycelium of branching, multinucleate, coenocytic hyphae.

Sporangiospores are shed from mature sporangia and dispersed by air currents or insects.

Sporangia develop at tips of vertical aerial hyphae. Cytoplasm and nuclei flow into the tip and divide up into segments to form sporangiospores (aplanospores).

Sexual cycle

Many members of the genus *Mucor* are homothallic and therefore each mycelium is self-fertile. *Mucor mucedo* is heterothallic, and sexual reproduction will only take place if two physiologically different and compatible mycelia (+ strain and – strain) are present together.

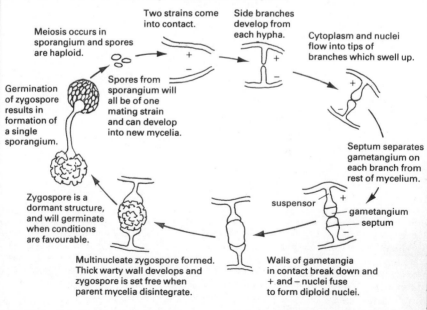

Meiosis occurs in sporangium and spores are haploid.

Two strains come into contact.

Side branches develop from each hypha.

Cytoplasm and nuclei flow into tips of branches which swell up.

Germination of zygospore results in formation of a single sporangium.

Spores from sporangium will all be of one mating strain and can develop into new mycelia.

Septum separates gametangium on each branch from rest of mycelium.

suspensor

gametangium
septum

Zygospore is a dormant structure, and will germinate when conditions are favourable.

Multinucleate zygospore formed. Thick warty wall develops and zygospore is set free when parent mycelia disintegrate.

Walls of gametangia in contact break down and + and – nuclei fuse to form diploid nuclei.

Rhizopus nigricans
Bread mould

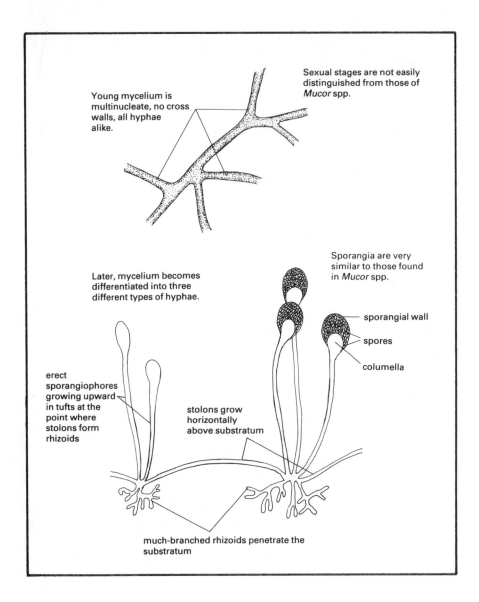

Young mycelium is multinucleate, no cross walls, all hyphae alike.

Sexual stages are not easily distinguished from those of *Mucor* spp.

Later, mycelium becomes differentiated into three different types of hyphae.

Sporangia are very similar to those found in *Mucor* spp.

sporangial wall

spores

columella

erect sporangiophores growing upward in tufts at the point where stolons form rhizoids

stolons grow horizontally above substratum

much-branched rhizoids penetrate the substratum

The genus *Rhizopus* contains about 30 species, all of which are saprophytic. *Rhizopus nigricans* commonly occurs on bread, and is a frequent contaminant of laboratory cultures. The structure and life history closely resemble those of *Mucor* spp. already described, except for the presence of stolons and rhizoids.

Class Ascomycetes

Most members of this class are terrestrial, and either parasitic or saprophytic. There are many parasitic species causing a wide variety of diseases in crop plants. The mycelia of most parasitic species grow within the host tissues, but in the powdery mildews the mycelium grows superficially. The saprophytic forms are found in a great variety of habitats, including soil, leaf mould and rotting wood, and they are often conspicuous because of their size and characteristic shapes. Some species may start their life as parasites, but later feed saprophytically on the dead tissues of the host.

The mycelium is septate in most species, but the yeasts are unicellular. Septa are incomplete walls with central perforations, allowing cytoplasmic streaming and the movement of nuclei from cell to cell. The cells are usually uninucleate and the hyphal walls contain a large proportion of chitin. The mycelium may be organized into tissues where the hyphae are either loosely compacted, as in prosenchyma, or closely compacted forming pseudoparenchyma. Such tissues are often associated with fruiting or spore-bearing structures. No motile cells are formed by any species in this class.

Asexual reproduction is by means of conidiospores, pycnospores, oidia or chlamydospores; fission or budding may occur, especially in the yeasts. Conidiospores are formed in succession on hyphae called conidiophores, which may arise separately from the mycelium, and be branched or unbranched. In some genera, the conidiophores are organized into definite fruiting bodies, the commonest of which is the pycnidium, a hollow, flask-shaped structure. Inside the pycnidium, pycnospores are produced from the lining of conidiophores. When many spores are formed simultaneously within a hypha, they are called oidia, and the single, large, thick-walled spores which sometimes occur are chlamydospores.

Sexual reproduction is by fusion of two compatible nuclei, which are brought together in a number of different ways. In some genera, definite male gametangia, the antheridia, and female ascogonia are formed, but in others the antheridia may be absent. Here, fusion of somatic hyphae occurs, or similar gametangia fuse in a manner which resembles that found in some of the Phycomycetes. Homothallic and heterothallic species occur. In most genera, the bringing together of compatible nuclei stimulates the ascogonium to produce a number of hyphal extensions, the ascogenous hyphae, into which pairs of nuclei migrate and undergo mitotic divisions. The ascogenous hyphae become septate and fusion of the compatible nuclei takes place in an ascus mother cell, which subsequently forms an ascus. The diploid zygote nucleus will immediately undergo meiosis to produce four haploid nuclei, each of which divides mitotically. The resulting eight haploid nuclei develop into ascospores, characteristic of the class. In most genera, the asci are produced in some kind of fruiting structure called an ascocarp, and release of the ascospores is achieved by forcible ejection from the ascus, the details varying according to the species.

Two sub-classes are recognized, according to presence or absence of ascogenous hyphae and ascocarps in the life cycle.

Sub-class Hemiascomycetes: no ascogenous hyphae or ascocarps present.
    Order Endomycetales   *Saccharomyces cerevisiae*

Sub-class Euascomycetes: ascogenous hyphae and ascocarps present.
    Order Plectascales   *Penicillium* spp.
                              Aspergillus spp.
    Order Erysiphales   *Sphaerotheca pannosa*
    Order Sphaeriales   *Neurospora crassa*
    Order Pezizales   *Bulgaria globosa*
                              Peziza spp.
                              Morchella esculenta

Order Endomycetales

Members of this order are mainly saprophytic, but a few parasitic species occur. The saprophytic species are abundant on substances which contain sugars, such as nectar and the surfaces of fruits. They also occur in soil, on animal faeces, in milk and the surface of plants. This order contains the yeasts, which are important economically for their ability to ferment sugars, yielding alcohol and carbon dioxide. This property is utilized in the baking and brewing industries, and special strains of yeasts have been developed for such purposes. Wild yeasts are still important in making wine.

The mycelium is often reduced, and many species are unicellular. A few saprophytic forms have a typical mycelium, but most species are unicellular. Cells may be spherical, ovoid or cylindrical, with a definite cell wall, a nucleus and often a large nuclear vacuole. The cytoplasm contains volutin granules, which provide reserves of phosphate, and glycogen. In the yeasts, chains of cells may be produced by budding.

Asexual reproduction is by means of fission or budding. Fission involves transverse division of a parent cell to give two equal daughter cells. The parent cell elongates, the nucleus divides and a transverse wall is laid down. In the budding process, a small outgrowth develops from the parent cell near one pole. The nucleus divides mitotically and one daughter nucleus migrates into the developing bud. A constriction divides the smaller daughter cell, or bud, from the parent. The daughter cell enlarges and may, in its turn, begin to bud before being separated from the original cell.

Sexual reproduction is by fusion of two vegetative cells, two ascospores or two gametangia. Amongst the yeasts, there is some variation in the sexual process from species to species, and even within the same species depending on conditions. After fusion, a diploid zygote is formed which develops into an ascus containing ascospores, the number varying from one to eight per ascus. The ascospores are released and give rise, by budding or fission, to vegetative cells.

Saccharomyces cerevisiae
Baker's or brewer's yeast

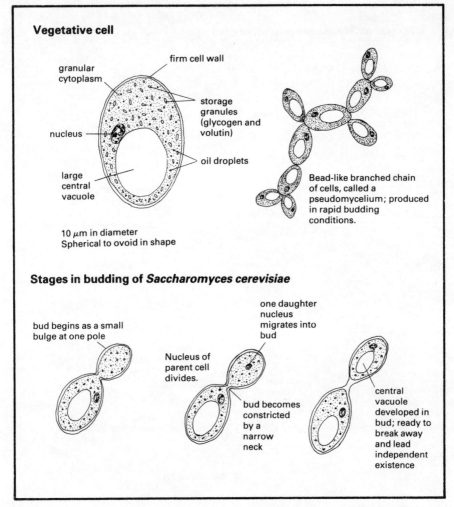

Vegetative cell

- firm cell wall
- granular cytoplasm
- storage granules (glycogen and volutin)
- nucleus
- oil droplets
- large central vacuole

10 μm in diameter
Spherical to ovoid in shape

Bead-like branched chain of cells, called a pseudomycelium; produced in rapid budding conditions.

Stages in budding of *Saccharomyces cerevisiae*

- bud begins as a small bulge at one pole
- Nucleus of parent cell divides.
- one daughter nucleus migrates into bud
- bud becomes constricted by a narrow neck
- central vacuole developed in bud; ready to break away and lead independent existence

Saccharomyces cerevisiae has a number of strains which are cultivated for specific purposes such as baking and brewing. In aerobic conditions, and with plentiful nutrients, *Saccharomyces* will grow and bud rapidly, but it is also able to survive in low oxygen concentrations and will undergo fermentation. In this latter process, sugars are incompletely broken down yielding carbon dioxide and alcohol, with no budding occurring under these conditions. When used in baking, the yeast's ability to produce carbon dioxide is used to make the bread dough spongy, and it is likely that some aerobic as well as anaerobic respiration will take place in the dough. When used in brewing, the yeast is used to ferment the sugars extracted from germinating barley, and alcohol is produced.

87

Saccharomyces cerevisiae
Baker's or brewer's yeast

DIVISION EUMYCOPHYTA

CLASS ASCOMYCETES

SUB-CLASS HEMIASCOMYCETES

ORDER ENDOMYCETALES

GENUS *Saccharomyces*

Life cycle of *Saccharomyces cerevisiae*

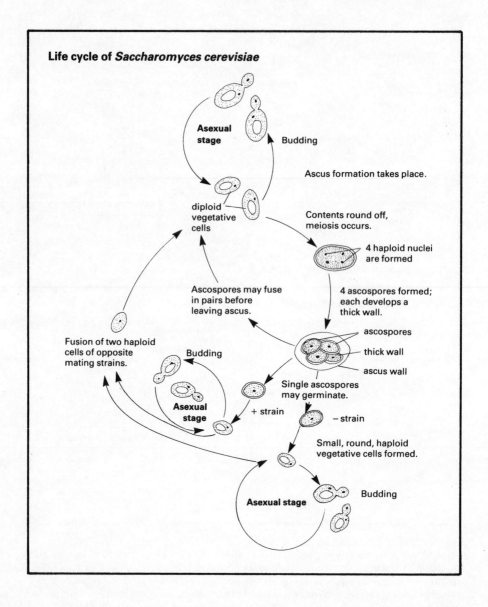

Asexual stage

Budding

Ascus formation takes place.

diploid vegetative cells

Contents round off, meiosis occurs.

4 haploid nuclei are formed

4 ascospores formed; each develops a thick wall.

Ascospores may fuse in pairs before leaving ascus.

ascospores

thick wall

ascus wall

Fusion of two haploid cells of opposite mating strains.

Budding

Asexual stage

+ strain

Single ascospores may germinate.

− strain

Small, round, haploid vegetative cells formed.

Asexual stage

Budding

Order Plectascales

Members of this order are mostly saprophytic, but a few are parasitic on plants and animals. This is a large order, and contains many common widespread moulds. The black, green and blue moulds on leather, fruit and preserves are all members of this order. Some species of *Aspergillus* are pathogenic and can cause diseases of the lungs in birds, cattle, sheep, horses and man, the symptoms resembling those of tuberculosis. Some members of this order are of economic importance and are used in the manufacture of organic acids, enzyme preparations and antibiotics. Some *Penicillium* species are important in the ripening of certain cheeses, and selected strains of *Penicillium notatum* and *Penicillium chrysogenum* are used for the large-scale production of the antibiotic penicillin.

The mycelium is septate and branched with multinucleate cells.

Asexual reproduction is by means of conidiospores. The young mycelium will produce many long, erect conidiophores. Conidiospores are produced in succession from special branches, or sterigmata, at the ends of the conidiophores, and dispersed by air currents. They will germinate rapidly on reaching a suitable substratum. This method of reproduction results in the formation of vast numbers of spores, and these fungi are common contaminants of exposed food.

Sexual reproduction occurs by the fusion of two compatible nuclei, which may be brought together in a number of different ways. The sexual or 'perfect' stages of many of the species in this order have never been observed, and it seems likely that these species have lost the ability to reproduce in this way. Where the perfect stage does occur, the ascocarp, which arises on a loose weft of hyphae, does not usually have an opening and is known as a cleistothecium. The wall may be formed of loosely-woven hyphae in some species, but of more compacted hyphae in the advanced types. Inside the cleistothecium, the asci are scattered at all levels; there is no orientation of the asci into a definite layer, or hymenium. The asci may be globose or club-shaped and many are stalked. No slit or pore develops in the asci; the ascospores are released by the deliquescence of the ascus tip or by breaking of the ascus. When the ascospores germinate, a germ tube is produced from which a new mycelium develops.

Aspergillus spp.

DIVISION EUMYCOPHYTA

CLASS ASCOMYCETES

SUB-CLASS EUASCOMYCETES

ORDER PLECTASCALES (ASPERGILLALES)

GENUS *Aspergillus*

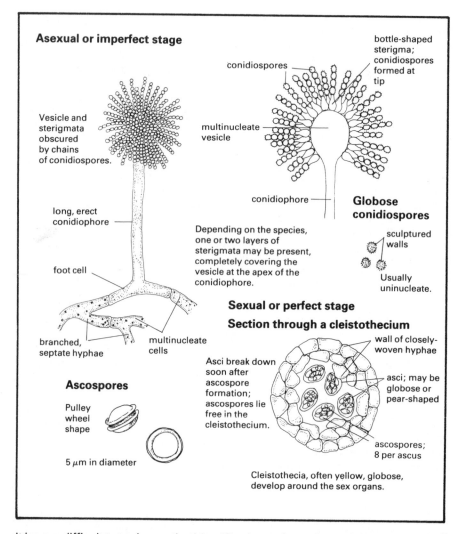

Asexual or imperfect stage

bottle-shaped sterigma; conidiospores formed at tip

conidiospores

Vesicle and sterigmata obscured by chains of conidiospores.

multinucleate vesicle

long, erect conidiophore

conidiophore

Globose conidiospores

Depending on the species, one or two layers of sterigmata may be present, completely covering the vesicle at the apex of the conidiophore.

sculptured walls

Usually uninucleate.

foot cell

branched, septate hyphae

multinucleate cells

Sexual or perfect stage

Section through a cleistothecium

wall of closely-woven hyphae

Ascospores

Pulley wheel shape

5 μm in diameter

Asci break down soon after ascospore formation; ascospores lie free in the cleistothecium.

asci; may be globose or pear-shaped

ascospores; 8 per ascus

Cleistothecia, often yellow, globose, develop around the sex organs.

It is very difficult to make precise identifications of members of this genus. Amongst the commonest saprophytes are the *Aspergillus glaucus* group, which form greenish-blue moulds on foods, and the *Aspergillus niger* group, causing the black moulds.

The colour of the conidiospores of an *Aspergillus* colony is a useful criterion for identification. The colonies may appear black, brown, green, blue or yellow depending on the species and the substratum on which the fungus is growing; thus it is possible to identify a particular species by culturing it on a medium of known nutrient composition.

Aspergillus spp.

Life cycle of *Aspergillus* spp.

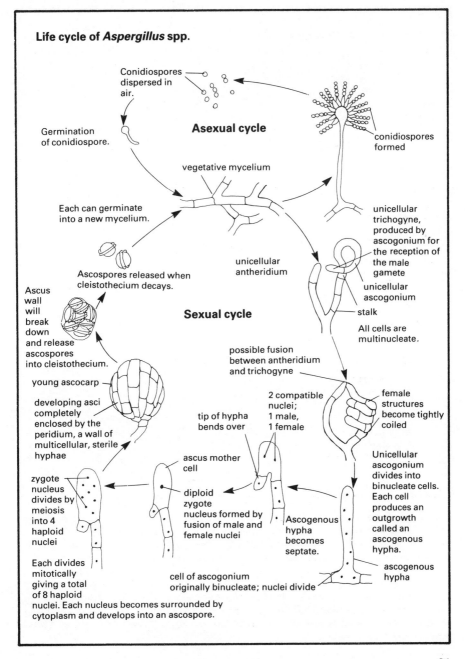

Conidiospores dispersed in air.

Asexual cycle

conidiospores formed

Germination of conidiospore.

vegetative mycelium

unicellular trichogyne, produced by ascogonium for the reception of the male gamete

Each can germinate into a new mycelium.

unicellular antheridium

unicellular ascogonium

Ascospores released when cleistothecium decays.

stalk

All cells are multinucleate.

Ascus wall will break down and release ascospores into cleistothecium.

Sexual cycle

possible fusion between antheridium and trichogyne

young ascocarp

female structures become tightly coiled

developing asci completely enclosed by the peridium, a wall of multicellular, sterile hyphae

2 compatible nuclei; 1 male, 1 female

tip of hypha bends over

Unicellular ascogonium divides into binucleate cells. Each cell produces an outgrowth called an ascogenous hypha.

ascus mother cell

zygote nucleus divides by meiosis into 4 haploid nuclei

diploid zygote nucleus formed by fusion of male and female nuclei

Ascogenous hypha becomes septate.

ascogenous hypha

Each divides mitotically giving a total of 8 haploid nuclei. Each nucleus becomes surrounded by cytoplasm and develops into an ascospore.

cell of ascogonium originally binucleate; nuclei divide

Penicillium spp.

DIVISION EUMYCOPHYTA

CLASS ASCOMYCETES

SUB-CLASS EUASCOMYCETES

ORDER PLECTASCALES (ASPERGILLALES)

GENUS *Penicillium*

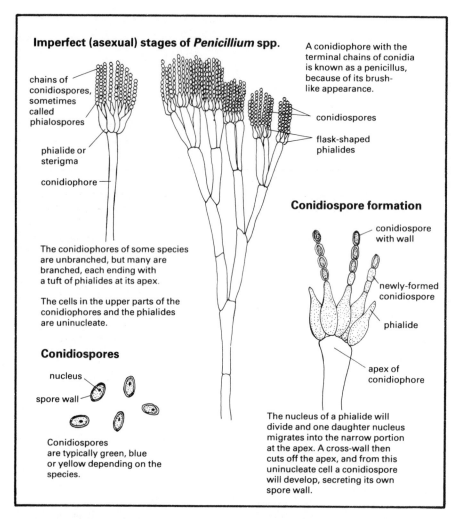

Imperfect (asexual) stages of *Penicillium* spp.

chains of conidiospores, sometimes called phialospores

phialide or sterigma

conidiophore

A conidiophore with the terminal chains of conidia is known as a penicillus, because of its brush-like appearance.

conidiospores

flask-shaped phialides

The conidiophores of some species are unbranched, but many are branched, each ending with a tuft of phialides at its apex.

The cells in the upper parts of the conidiophores and the phialides are uninucleate.

Conidiospores

nucleus

spore wall

Conidiospores are typically green, blue or yellow depending on the species.

Conidiospore formation

conidiospore with wall

newly-formed conidiospore

phialide

apex of conidiophore

The nucleus of a phialide will divide and one daughter nucleus migrates into the narrow portion at the apex. A cross-wall then cuts off the apex, and from this uninucleate cell a conidiospore will develop, secreting its own spore wall.

Members of the genus *Penicillium* are mostly saprophytic on decaying fruit and vegetables; some can survive in conditions where there is little moisture. The genus is closely related to *Aspergillus*, with the imperfect stages in the life cycle predominating. However, the conidiophores of *Penicillium* lack foot cells and do not develop vesicles, but often possess a compact series of branches from which the phialides, or sterigmata, arise and produce the conidiospores.

A few species are known to reproduce sexually, but the vast majority only show asexual reproduction. Where cleistothecia are produced, they closely resemble those of *Aspergillus* and the details of ascus formation and ascospore release are very similar.

Order Erysiphales

Members of this order are mainly obligate parasites, externally parasitic on the leaves and stems of Angiosperms. They are responsible for the group of plant diseases known as the powdery mildews which are of economic importance, causing complete destruction of affected crops in some regions.

The mycelium is branched and septate, with short, uninucleate cells. It spreads over the surface of the host, producing one-celled haustorial branches which penetrate into the epidermal cells and absorb nutrients.

Asexual reproduction is by conidiospores and takes place on young mycelia. A large number of short, upright unicellular conidiophores grow from the mycelium. Conidiospores may be cut off in succession from the tip, or the conidiophore may be divided into a long stalk cell and a short cell at the apex, from which the conidiospores are cut off. Conidiospores are uninucleate and oval, produced in large numbers and germinate immediately they land on a suitable host. The new mycelium formed will be capable of producing conidiospores within a few days.

Sexual reproduction is by means of well-defined sex organs or by vegetative cells which serve the same function. In most species, definite antheridia and ascogonia are formed, and after the ascogonium becomes dikaryotic, a globose cleistothecium with a pseudoparenchymatous peridium develops. Usually the asci are arranged in a single layer at the base of the cleistothecium, but in some genera only a single ascus develops. Mature cleistothecia, or cleistocarps, may remain attached to the host, or become detached and dispersed by the wind. The cleistothecia, with their thick peridia (there may be 6 to 10 layers of cells) usually survive the winter and split open in the following spring. The ascospores may be forcibly ejected and will germinate immediately they land on a suitable host.

Sphaerotheca pannosa

DIVISION EUMYCOPHYTA

CLASS ASCOMYCETES

SUB-CLASS EUASCOMYCETES

ORDER ERYSIPHALES

GENUS *Sphaerotheca*

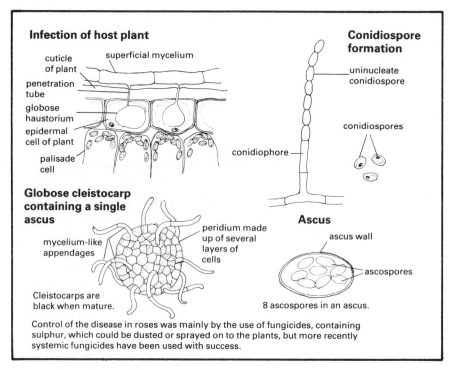

Infection of host plant

- cuticle of plant
- superficial mycelium
- penetration tube
- globose haustorium
- epidermal cell of plant
- palisade cell

Conidiospore formation

- uninucleate conidiospore
- conidiospores
- conidiophore

Globose cleistocarp containing a single ascus

- mycelium-like appendages
- peridium made up of several layers of cells

Cleistocarps are black when mature.

Ascus

- ascus wall
- ascospores

8 ascospores in an ascus.

Control of the disease in roses was mainly by the use of fungicides, containing sulphur, which could be dusted or sprayed on to the plants, but more recently systemic fungicides have been used with success.

This fungus causes powdery mildew of roses. In the spring and early summer, the aerial parts of the rose plant will be infected by conidiospores. These germinate to produce a mycelium which sends fine branches through the cuticle and into the epidermal cells of the leaves. Globose haustoria are formed and absorb nutrients from the epidermal cells. After about four days, the superficial mycelium produces conidiophores which begin to cut off conidiospores. The conidiospores are easily detached and will germinate readily in the right conditions.

This cycle of infection and sporulation can occur throughout the growing season of the host, and can be quite severe, seriously reducing the photosynthetic activities of the host. In this species, the germination of the conidiospores seems to be favoured by high humidity, thus the disease will flourish in damp, warm conditions.

In late summer, globose cleistothecia, or cleistocarps, containing a single ascus develop on the superficial mycelium. The cleistocarps have a very thick peridium and are black when mature. It is possible that the cleistocarps overwinter, and that the single ascus is released the following spring, but it is thought unlikely that the ascospores provide the only inoculum for infection the following season. The mycelium becomes established in the buds, and when they grow in the spring, they produce mildewed shoots which can cause a very rapid spread of infection.

Order Sphaeriales

Members of this order are mostly saprophytic, but some are parasitic on vascular plants. This is a large order, containing several thousand species. The saprophytic species grow on dead twigs, the branches, leaves and stems of plants, or on paper and cloth, and there are some species which grow on dung (coprophilous). The species which are capable of breaking down cellulose are the greatest nuisance to man, causing damage to timber and other fabrics. One well-known saprophytic genus is *Neurospora*, which causes red mould of bread. It has long been a pest in bakeries because it produces masses of conidiospores and its hyphae grow very rapidly. This genus has been used extensively in the study of fungal genetics. Amongst the parasitic species is *Ophiostoma ulmi*, the cause of Dutch elm disease which has spread so rapidly resulting in the death of large numbers of elm trees in Britain in the last few years.

The mycelium is freely branching, with uninucleate or multinucleate, elongated cells. Asexual reproduction is by conidiospores. Conidiospores are produced on conidiophores which may be enclosed in a fruiting structure, or organized into groups, or show no organization at all. The conidiospores may be coloured or hyaline, and vary in size and shape according to the species. A very wide range of conidiospore characteristics occurs in this order. Some species produce spermatia, which are uninucleate spore-like bodies, budding off laterally from the cells of an erect structure resembling a conidiophore. These spermatia are often called microconidia because of their small size, to distinguish them from the conidiospores, or macroconidia. The spermatia can function as conidiospores, germinating to form new mycelia.

Sexual reproduction is usually by means of spermatia or conidia uniting with trichogynes from ascogonia. Antheridia and ascogonia are present in many species of this order, but the antheridia do not function. Each ascogonium becomes multicellular and sterile hyphae grow round this to form a globose envelope. Several of the ascogonial cells develop trichogynes, which will unite with spermatia or conidia near them, the nuclei from the spermatia or conidia migrating into the trichogynes. Shortly after spermatization, the ascogonium will produce bi-nucleate ascogenous hyphae from which asci develop. Formation of the fruiting body or perithecium involves growth of a pseudoparenchymatous peridium, the upper portion of which becomes a long neck terminating in a pore. In many species, the perithecia have sterile hairs, paraphyses, amongst the asci, and periphyses just beneath the pore, or ostiole.

Neurospora crassa

DIVISION EUMYCOPHYTA

CLASS ASCOMYCETES

SUB-CLASS EUASCOMYCETES

ORDER SPHAERIALES

GENUS Neurospora

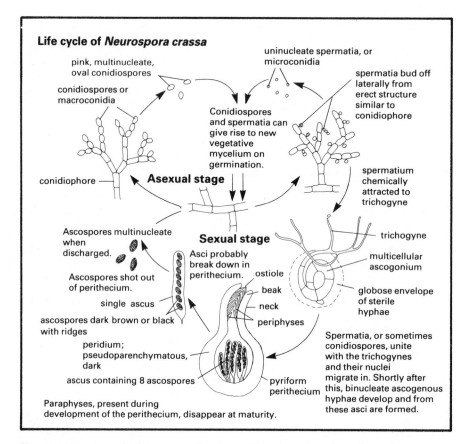

Life cycle of *Neurospora crassa*

pink, multinucleate, oval conidiospores

conidiospores or macroconidia

conidiophore

uninucleate spermatia, or microconidia

spermatia bud off laterally from erect structure similar to conidiophore

Conidiospores and spermatia can give rise to new vegetative mycelium on germination.

Asexual stage

spermatium chemically attracted to trichogyne

Ascospores multinucleate when discharged.

Ascospores shot out of perithecium.

single ascus

ascospores dark brown or black with ridges

peridium; pseudoparenchymatous, dark

ascus containing 8 ascospores

Paraphyses, present during development of the perithecium, disappear at maturity.

Sexual stage

Asci probably break down in perithecium.

ostiole

beak

neck

periphyses

pyriform perithecium

trichogyne

multicellular ascogonium

globose envelope of sterile hyphae

Spermatia, or sometimes conidiospores, unite with the trichogynes and their nuclei migrate in. Shortly after this, binucleate ascogenous hyphae develop and from these asci are formed.

Neurospora crassa is the cause of red mould on bread and produces large numbers of pink conidiospores during the asexual stage of its life cycle. This is the stage most commonly found. In order for sexual reproduction to take place, two strains of opposite mating type have to be present, and spermatia or conidiospores act as male cells in the absence of antheridia. Eight ascospores are formed in each ascus in this species, four of one mating type and four of the other.

Beadle and Tatum, working at Stanford University, California, used *Neurospora crassa* in experiments to support their one-gene-one-enzyme hypothesis. They showed that the mould would thrive on a basic medium, synthesizing all its own amino acids. Using X-rays, Beadle and Tatum produced mutants which could not grow on the basic medium because they had lost the ability to synthesize a particular amino acid. They then showed that this trait could be inherited in Mendelian fashion, which suggested that the synthesis of the amino acid was due to the presence of one enzyme controlled by a single gene.

Order Pezizales

Members of this order are mostly saprophytic. Many species live in the soil and form their conspicuous, often brightly coloured, fruiting structures, or apothecia, above ground. Some may be found on dead wood, humus or dung, and the apothecia can be found from spring until late autumn. They vary in size from less than 1 mm to several cm in diameter. The members of this order are of little economic importance to man; there are a few edible species of *Morchella*, the morels, and the genus *Pyronema* occurs frequently on burnt ground, or soil that has been sterilized by heating or steaming.

The mycelium is branched and septate, with multinucleate cells.

Asexual reproduction is by conidiospores or oidia. The imperfect stages of most of the members of this order are not known, but conidiospore formation has been reported in several species of *Peziza*. Microconidia, or spermatia, are known to be produced in some species, and in others erect hyphae may form chains of oidia.

Sexual reproduction is similar to that found in other members of the sub-class; antheridia and ascogonia are formed in some species. The distinguishing features of sexual reproduction in this order are the form of the ascocarp and the method of liberation of the ascospores. The ascocarp is typically an open cup- or disc-shaped structure, the apothecium, bearing the asci in a layer on the surface, intermingled with paraphyses. The ascospores are liberated from the ascus through an opening formed by a definite lid or operculum.

Some Pezizales

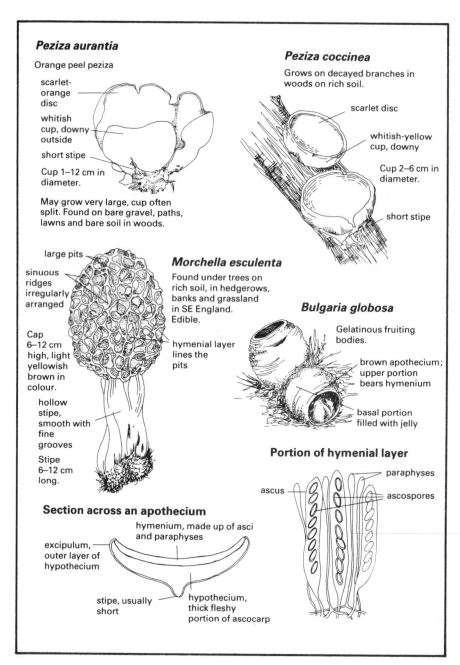

Peziza aurantia

Orange peel peziza

scarlet-orange disc

whitish cup, downy outside

short stipe

Cup 1–12 cm in diameter.

May grow very large, cup often split. Found on bare gravel, paths, lawns and bare soil in woods.

Peziza coccinea

Grows on decayed branches in woods on rich soil.

scarlet disc

whitish-yellow cup, downy

Cup 2–6 cm in diameter.

short stipe

Morchella esculenta

large pits

sinuous ridges irregularly arranged

Cap 6–12 cm high, light yellowish brown in colour.

hymenial layer lines the pits

Found under trees on rich soil, in hedgerows, banks and grassland in SE England. Edible.

hollow stipe, smooth with fine grooves

Stipe 6–12 cm long.

Bulgaria globosa

Gelatinous fruiting bodies.

brown apothecium; upper portion bears hymenium

basal portion filled with jelly

Portion of hymenial layer

paraphyses

ascus

ascospores

Section across an apothecium

hymenium, made up of asci and paraphyses

excipulum, outer layer of hypothecium

stipe, usually short

hypothecium, thick fleshy portion of ascocarp

Class Basidiomycetes

The members of this class are terrestrial, and saprophytic or parasitic. This is an important group of fungi, containing parasitic species of economic importance to man. The Ustilaginales, smut fungi, and the Uredinales, rust fungi, cause serious diseases of cereal crops, and amongst the higher Basidiomycetes there are many which infect forest trees and destroy timber. *Serpula lacrymans* causes extensive damage to houses, infecting damp timber and giving rise to the condition known as 'dry rot'. Many species are saprophytic on rotten logs, wet leaves and other organic matter, including dung. Some species are sought after as food, and many form mycorrhizal associations with higher plants.

The mycelium is septate, often showing clamp connections and cross-connections or anastomoses. Two types of mycelium are found: a primary mycelium made up of uninucleate cells, which usually develops from the germination of a basidiospore and on which basidia never form; and a secondary mycelium of binucleate cells arising by fusion of two uninucleate cells. In heterothallic species, this fusion of cells occurs when primary mycelia of opposite sex grow together. In homothallic species, the fusion may occur between two hyphae of a single primary mycelium. Secondary mycelia may also arise from the fusion of a primary mycelium with an oidium or spermatium from another primary mycelium. The binucleate cells of the secondary mycelium divide to produce daughter cells, the two nuclei dividing simultaneously.

In most genera, special structures called clamp connections are formed and these ensure that sister nuclei are separated during the process of daughter cell formation. Complex fruiting bodies, or sporophores, are developed in the more advanced types, where the hyphae are organized into structures of characteristic shape and often large size. In some species, hyphae may lie parallel to one another, forming thick strands of mycelium called rhizomorphs.

Asexual reproduction is by conidiospores, oidia, budding or by fragmentation of the mycelium. Conidiospore production is commonly found in the Uredinales and Ustilaginales. In many species, the hyphae of both primary and secondary mycelia may break up into unicellular sections, which produce germ tubes and develop into new mycelia. Oidia are cut off in succession from special branches of the hyphae, the oidiophores, and may either germinate to produce new mycelia or may fuse with hyphae from a primary mycelium to give a secondary mycelium.

Sexual reproduction is by fusion of nuclei in a basidium; no specialized sex organs are formed. Basidia only develop on the dikaryotic secondary mycelium and may be septate (phragmobasidia) or non-septate (holobasidia). After fusion of the two haploid nuclei in the basidium, the diploid zygote nucleus undergoes meiosis to give rise to four haploid nuclei, which migrate into the swelling sterigmata at the apex of the basidium. A cross wall cuts off each basidiospore from the basidium. Many basidiospores are discharged quite violently from the basidia, and will germinate immediately in suitable conditions to produce a primary mycelium.

Two sub-classes are recognized.

Sub-class Heterobasidiomycetes: basidia are septate (phragmobasidia).
    Order Uredinales   *Puccinia* spp.

Sub-class Homobasidiomycetes: basidia have no septa (holobasidia).
    Order Agaricales   *Agaricus*
                             Coprinus
    Order Aphyllophorales   *Polyporus* (*Piptoporus*)
                             Serpula lacrymans

Order Uredinales

All members of this order are obligate parasites of vascular plants. Amongst the diseases caused by fungi in this order are rusts of coffee, asparagus, beans, carnations and cereals. Many of these fungi are widespread, occurring where susceptible host plants are found, and the cereal rusts have been known since Roman times.

The mycelium is uninucleate at first, becoming binucleate later. The mycelium grows intercellularly in the host plant, penetrating the host cells by haustoria, and absorbing nutrients. Clamp connections may be found in the binucleate mycelium of some species.

Reproduction is complicated, with a variety of spores produced at different stages, and the members of this order can be divided into two groups according to the length and complexity of the life cycle. In the first group, the microcyclic rusts, the life cycle is short, with only two types of spore produced: teleutospores and basidiospores. Some species may produce spermatia. Basidiospores each contain a haploid nucleus and germinate to produce a primary mycelium which grows intercellularly in the host plant. The primary mycelium becomes dikaryotic by fusion of hyphae and will then form a teleutosorus and produce dikaryotic teleutospores from which basidia and basidiospores develop.

   In the second group, the macrocyclic rusts, up to three different types of dikaryotic spores may be produced, together with basidiospores and spermatia. In this group, two distinct hosts are often necessary for the complete life cycle to take place. As in the microcyclic rusts, dikaryotic teleutospores are formed on one host, and after fusion of the nuclei, basidia and basidiospores develop. The basidiospores need to infect a new host, where they germinate and give rise to a primary mycelium, which soon produces spermogonia. Inside the flask-shaped spermogonia, or pycnidia, uninucleate spermatia, or pycnospores, are cut off, and receptive hyphae emerge from the pore, or ostiole. Spermatia are extruded in a drop of sugary fluid and transferred by flies to the receptive hyphae of a spermogonium of the opposite mating strain, where spermatization takes place and a binucleate mycelium develops.

   The primary mycelium may also produce proto-aecidia, which undergo no further development until the mycelium becomes binucleate after spermatization. If this happens, then the proto-aecidia become aecidia, and produce dikaryotic aecidiospores which can infect a different host. After infection, an asexual stage involving the formation of a third type of dikaryotic spore, the uredospore, then occurs, before teleutospore production begins.

Puccinia graminis

DIVISION EUMYCOPHYTA

CLASS BASIDIOMYCETES

SUB-CLASS HETEROBASIDIOMYCETES

ORDER UREDINALES

GENUS *Puccinia*

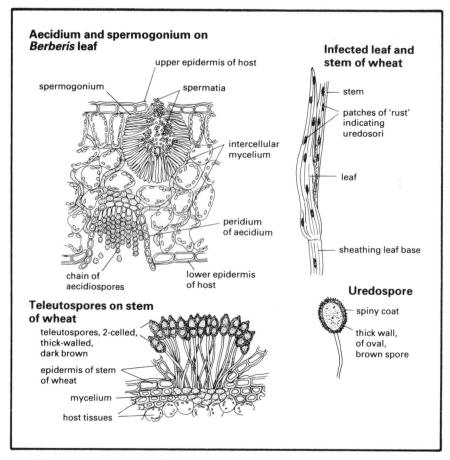

Aecidium and spermogonium on *Berberis* leaf

upper epidermis of host

spermogonium

spermatia

intercellular mycelium

peridium of aecidium

chain of aecidiospores

lower epidermis of host

Infected leaf and stem of wheat

stem

patches of 'rust' indicating uredosori

leaf

sheathing leaf base

Uredospore

spiny coat

thick wall, of oval, brown spore

Teleutospores on stem of wheat

teleutospores, 2-celled, thick-walled, dark brown

epidermis of stem of wheat

mycelium

host tissues

Puccinia graminis causes black stem rust of cereals, and there are a number of different sub-species which attack specific hosts, e.g. *P. graminis tritici* occurs on wheat, and *P. graminis avenae* occurs on oats but never on wheat. The basidiospores of all the sub-species can infect the barberry, *Berberis vulgaris*. *P. graminis* does not kill its hosts, and the main effect on the wheat is to reduce the size of the grains and thus the yield from the crop. Due to the great variety among the cultivated host plants, each sub-species of *Puccinia* contains a series of physiological races which vary in their ability to attack the varieties of host plant and cause infection.

Control can best be achieved by breeding disease-resistant varieties of wheat, but it has proved difficult because of the large number of physiological races of the fungus. Eradication of the second host plant, the barberry, and dusting with copper and sulphur fungicides have been tried with some success, but the latter can be impractical on a large scale.

DIVISION EUMYCOPHYTA
CLASS BASIDIOMYCETES
SUB-CLASS HETEROBASIDIOMYCETES
ORDER UREDINALES
GENUS *Puccinia*

Life cycle of *Puccinia graminis*

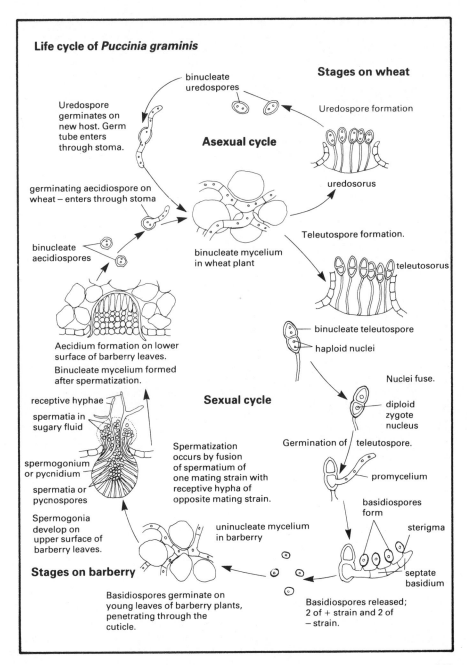

Stages on wheat

binucleate uredospores

Uredospore formation

Uredospore germinates on new host. Germ tube enters through stoma.

Asexual cycle

uredosorus

germinating aecidiospore on wheat – enters through stoma

Teleutospore formation.

teleutosorus

binucleate aecidiospores

binucleate mycelium in wheat plant

binucleate teleutospore

haploid nuclei

Nuclei fuse.

Aecidium formation on lower surface of barberry leaves. Binucleate mycelium formed after spermatization.

diploid zygote nucleus

receptive hyphae

Sexual cycle

Germination of teleutospore.

spermatia in sugary fluid

promycelium

spermogonium or pycnidium

Spermatization occurs by fusion of spermatium of one mating strain with receptive hypha of opposite mating strain.

basidiospores form

spermatia or pycnospores

sterigma

Spermogonia develop on upper surface of barberry leaves.

uninucleate mycelium in barberry

septate basidium

Stages on barberry

Basidiospores germinate on young leaves of barberry plants, penetrating through the cuticle.

Basidiospores released; 2 of + strain and 2 of – strain.

Order Agaricales

The members of this order are mostly saprophytic, but a few parasitic species exist. Saprophytic species grow on soil, or on dead and decaying wood and leaves. Many form mycorrhizal associations with forest trees, and this is why it is common to find certain species of agarics growing in specific types of woodland, e.g. the common fly agaric forms an association with the roots of birch trees. Many species are edible and some are sought-after for food, not for their nutritional value but for their special flavours. Some species are definitely poisonous, though many more have a disagreeable taste.

The mycelium is septate, and either uninucleate or binucleate. The primary mycelium forms a secondary mycelium, which is binucleate, by hyphal fusion. Many members of this order have clamp connections on the secondary mycelium. In some cases, many hyphae may lie parallel to one another, or twisted together, forming thick strands called rhizomorphs.

Asexual reproduction is rare. A few species produce conidiospores or oidia, but very few members of this order have been shown to undergo a distinct asexual reproductive stage. Some species of *Coprinus* produce oidia, and chlamydospores occur in *Agaricus* species.

Sexual reproduction is by means of hyphal fusion followed by the formation of basidia and basidiospores. The basidia form a hymenium on gills, or lamellae, which are borne in a radial arrangement on the undersurface of the fruiting body or sporophore. In the simple forms, the gills are exposed from the beginning, but in the more complex forms the developing gills are covered by a partial veil, which ruptures as the pileus enlarges, leaving a ring, or annulus, around the stipe. In some, there is a universal veil, which totally encloses the young sporophore, rupturing when the stipe elongates, leaving scales on the cap and a bag-like structure, the volva, round the base of the stipe. Most of the Agaricales are heterothallic.

DIVISION EUMYCOPHYTA
CLASS BASIDIOMYCETES
SUB-CLASS HOMOBASIDIOMYCETES
ORDER AGARICALES
GENUS *Agaricus*

Agaricus campestris

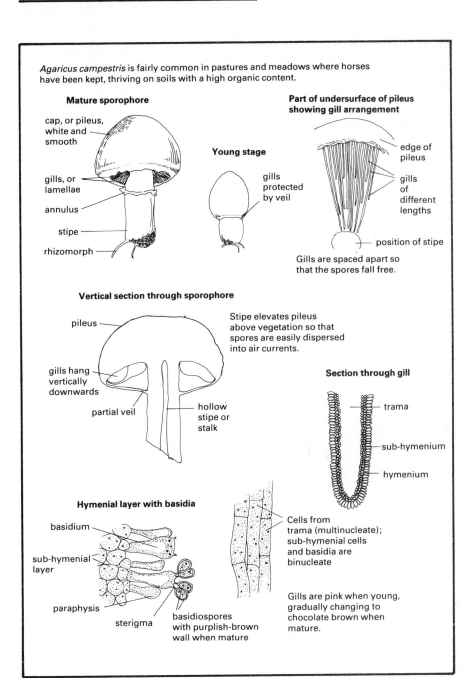

Agaricus campestris is fairly common in pastures and meadows where horses have been kept, thriving on soils with a high organic content.

Mature sporophore

cap, or pileus, white and smooth

gills, or lamellae

annulus

stipe

rhizomorph

Young stage

gills protected by veil

Part of undersurface of pileus showing gill arrangement

edge of pileus

gills of different lengths

position of stipe

Gills are spaced apart so that the spores fall free.

Vertical section through sporophore

pileus

gills hang vertically downwards

partial veil

hollow stipe or stalk

Stipe elevates pileus above vegetation so that spores are easily dispersed into air currents.

Section through gill

trama

sub-hymenium

hymenium

Hymenial layer with basidia

basidium

sub-hymenial layer

paraphysis

sterigma

basidiospores with purplish-brown wall when mature

Cells from trama (multinucleate); sub-hymenial cells and basidia are binucleate

Gills are pink when young, gradually changing to chocolate brown when mature.

Agaricus campestris

DIVISION EUMYCOPHYTA

CLASS BASIDIOMYCETES

SUB-CLASS HOMOBASIDIOMYCETES

ORDER AGARICALES

GENUS *Agaricus*

Life cycle of *Agaricus campestris*

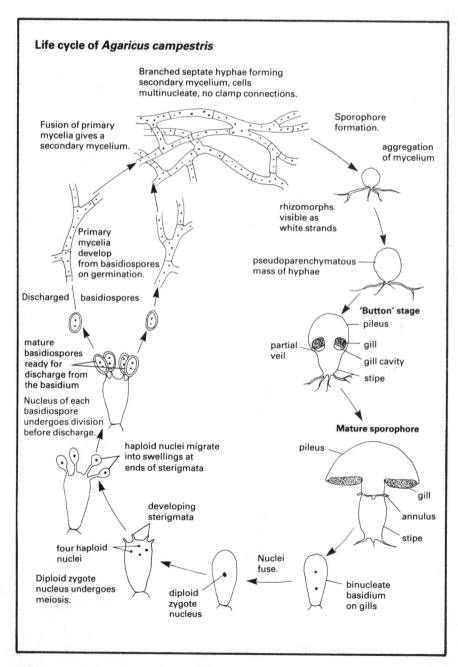

Branched septate hyphae forming secondary mycelium, cells multinucleate, no clamp connections.

Fusion of primary mycelia gives a secondary mycelium.

Sporophore formation.

aggregation of mycelium

Primary mycelia develop from basidiospores on germination.

rhizomorphs visible as white strands

Discharged | basidiospores

pseudoparenchymatous mass of hyphae

mature basidiospores ready for discharge from the basidium

'Button' stage
- pileus
- gill
- gill cavity
- stipe

partial veil

Nucleus of each basidiospore undergoes division before discharge.

haploid nuclei migrate into swellings at ends of sterigmata

Mature sporophore

pileus

gill

annulus

stipe

developing sterigmata

four haploid nuclei

Nuclei fuse.

Diploid zygote nucleus undergoes meiosis.

diploid zygote nucleus

binucleate basidium on gills

106

Coprinus spp.

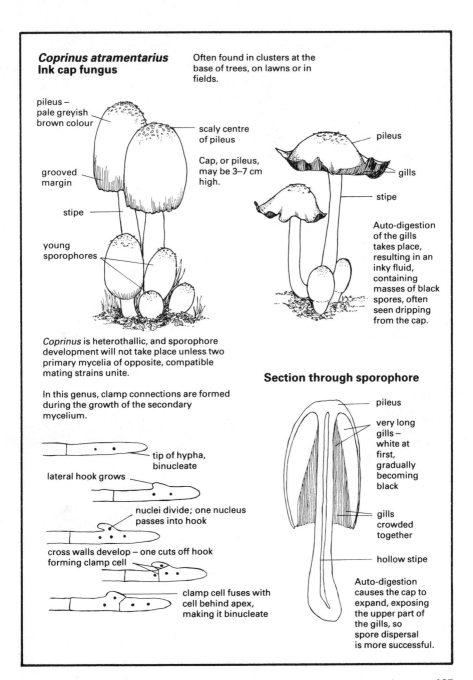

**Coprinus atramentarius
Ink cap fungus**

Often found in clusters at the base of trees, on lawns or in fields.

pileus –
pale greyish
brown colour

scaly centre
of pileus

Cap, or pileus,
may be 3–7 cm
high.

grooved
margin

stipe

young
sporophores

pileus

gills

stipe

Auto-digestion
of the gills
takes place,
resulting in an
inky fluid,
containing
masses of black
spores, often
seen dripping
from the cap.

Coprinus is heterothallic, and sporophore development will not take place unless two primary mycelia of opposite, compatible mating strains unite.

In this genus, clamp connections are formed during the growth of the secondary mycelium.

tip of hypha,
binucleate

lateral hook grows

nuclei divide; one nucleus
passes into hook

cross walls develop – one cuts off hook
forming clamp cell

clamp cell fuses with
cell behind apex,
making it binucleate

Section through sporophore

pileus

very long
gills –
white at
first,
gradually
becoming
black

gills
crowded
together

hollow stipe

Auto-digestion
causes the cap to
expand, exposing
the upper part of
the gills, so
spore dispersal
is more successful.

107

Order Aphyllophorales

This order contains many parasitic species, and a few saprophytes. Amongst the parasitic species are *Stereum hirsutum*, causing heart rot of oaks, *Stereum purpureum*, causing silver leaf disease of plums, and *Sparassis radicata* causing root and trunk disease in conifers. This group contains the bracket fungi which grow on woodland trees. The fungus mycelium is present in the soil and infects the host tree through the roots or through a wound. The mycelium then spreads through the phloem just below the bark, absorbing food from the living cells; some species can break down lignin and feed on the xylem tissue as well. Often fruiting bodies are not formed until the tree is dead. Saprophytic species occur on rotting wood, some causing considerable damage to stored timber.

The mycelium is septate and branched.

Asexual reproduction is by means of conidiospores or oidia. The conidiophores are spherical and bear clusters of spherical or oval conidiospores, which develop to produce a dikaryotic mycelium. Oidia are thick-walled, spherical spores formed directly from the mycelium, and are thought to give rise to the primary mycelium on germination.

Sexual reproduction is by fusion of compatible nuclei in a basidium followed by basidiospore formation. In fungi in this order, the hymenium is exposed as soon as the sporophore develops, and may be found on the surface of teeth, spines or gills, or lining narrow tubes.

Piptoporus betulinus

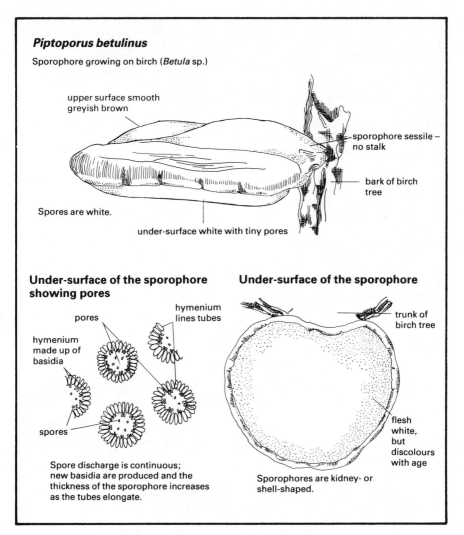

Piptoporus betulinus

Sporophore growing on birch (*Betula* sp.)

upper surface smooth greyish brown

sporophore sessile – no stalk

bark of birch tree

Spores are white.

under-surface white with tiny pores

Under-surface of the sporophore showing pores

pores

hymenium lines tubes

hymenium made up of basidia

spores

Spore discharge is continuous; new basidia are produced and the thickness of the sporophore increases as the tubes elongate.

Under-surface of the sporophore

trunk of birch tree

flesh white, but discolours with age

Sporophores are kidney- or shell-shaped.

Piptoporus betulinus (formerly *Polyporus betulinus*) is quite common in birch woods, and sporophores may be found on living trees as well as on fallen logs and trunks. The mycelium develops beneath the bark, forming a complete sheet of white tissue, which gets thicker as the parasite becomes established. The fruiting bodies begin as knobs of tissue, which push through the bark and gradually grow, the largest size achieved being about 30 cm in diameter. The fruiting bodies only last for a single season in this species.

The sporophores of *Piptoporus betulinus* have been used in the manufacture of charcoal crayons, and the upper surface makes a very good razor strop.

Lichens

The lichens are an interesting group of plants, each being made up of an alga and a fungus living together in a close association. The relationship is thought by some to be a symbiotic one, where both partners benefit, but others have suggested that the fungus benefits at the expense of the alga, so it should more properly be described as a parasitic one. Whatever the relationship, the union of the two components gives rise to thalloid structures which are considered to be distinct plants, and some 400 different genera have been described.

The fungal component (mycobiont) is an Ascomycete or a Basidiomycete, and the algal component (phycobiont) may belong to the Cyanophyta or the Chlorophyta. Fungal hyphae form the vegetative thallus and eventually the fruiting bodies, and the algae are usually limited to a thin layer just below the surface.

Lichens are found growing in a wide variety of situations. Some species are able to live where there is no other vegetation, and are important colonizers of bare rock or inhabitants of arctic regions, while others flourish on the leaves and bark of trees in tropical rain forests. Many lichens are extremely sensitive to atmospheric pollutants such as sulphur dioxide. It has been shown that the sulphur dioxide destroys the chlorophyll in the algal component, and, in high concentrations, eventually causes irreversible damage to the cells. Amongst other pollutants which have been shown to be harmful are fluorides and soot, and the lack of lichens in urban areas has been attributed to pollution plus the lower humidity and higher temperatures found there.

Lichens vary in their sensitivity to pollutants, and can be used to assay pollution levels. Indicator species, varying in their tolerance of specific pollutants, are recognized, and lichens are of value in both quantitative and qualitative monitoring of man-made pollution.

As the thalli and fruiting structures of the lichens are fungal in structure, no taxonomic significance is attached to the algal component, and the classification is based solely on the fungal characteristics. The lichens are assigned sub-division status in the Eumycophyta, and two classes are recognized.

Class Ascolichenes: fungal component is an Ascomycete.
   Examples: *Xanthoria*
                 Peltigera

Class Basidiolichenes: fungal component is a Basidiomycete.
   Only four genera described with no British representatives.

Xanthoria parietina

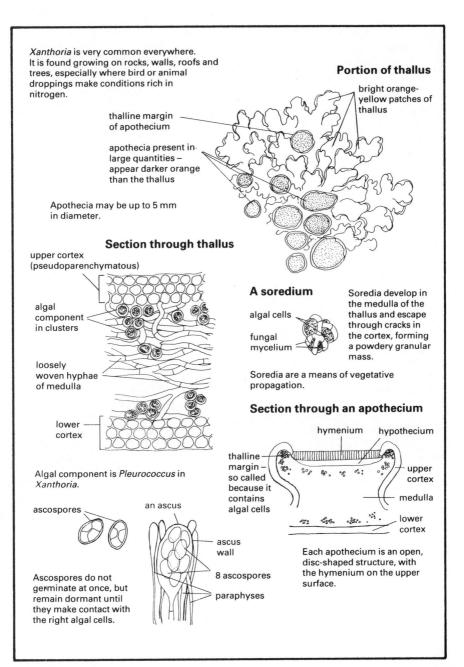

Xanthoria is very common everywhere. It is found growing on rocks, walls, roofs and trees, especially where bird or animal droppings make conditions rich in nitrogen.

Portion of thallus

bright orange-yellow patches of thallus

thalline margin of apothecium

apothecia present in large quantities – appear darker orange than the thallus

Apothecia may be up to 5 mm in diameter.

Section through thallus

upper cortex (pseudoparenchymatous)

algal component in clusters

loosely woven hyphae of medulla

lower cortex

Algal component is *Pleurococcus* in *Xanthoria*.

A soredium

algal cells

fungal mycelium

Soredia develop in the medulla of the thallus and escape through cracks in the cortex, forming a powdery granular mass.

Soredia are a means of vegetative propagation.

Section through an apothecium

hymenium   hypothecium

thalline margin – so called because it contains algal cells

upper cortex

medulla

lower cortex

Each apothecium is an open, disc-shaped structure, with the hymenium on the upper surface.

ascospores

an ascus

Ascospores do not germinate at once, but remain dormant until they make contact with the right algal cells.

ascus wall

8 ascospores

paraphyses

111

Lichens

DIVISION EUMYCOPHYTA

SUB-DIVISION LICHENES

CLASS ASCOLICHENES

GENUS *Peltigera*   *Cladonia*

Usnea   *Hypogymnia*

Peltigera canina

Common on walls, rocks and soil; also found amongst grass in woods, on lawns and sand-dunes.

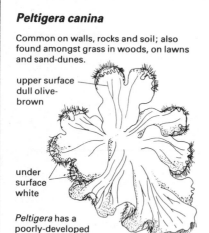

upper surface dull olive-brown

under surface white

Peltigera has a poorly-developed lower cortex covered with root-like growths. It forms large lobed patches.

Cladonia floerkeana

Common on moorland and in hilly areas.

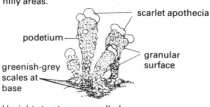

scarlet apothecia

podetium

granular surface

greenish-grey scales at base

Upright structures are called podetia, and are simple greyish stalks in this species. Apothecia develop on the apex of the podetia; no cups are formed.

Hypogymnia physodes (Parmelia physodes)

Common on twigs, branches, wood, rocks and walls, where it may form large, much-branched masses.

narrow smooth lobes, greenish-grey in colour

twig

Usnea subfloridana

bark of tree

grey-green, much-branched tufts 3–8 cm long

Common on trees, especially in hilly areas in the north of Britain.

Cladonia coccifera

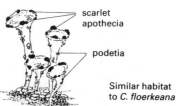

scarlet apothecia

podetia

Similar habitat to *C. floerkeana*

Podetia form wide-mouthed cups with apothecia on the rims.

Glossary

acronematic – describes 'whiplash' type of flagellum which is long and smooth

adventitious – describes organs which appear in uncharacteristic positions, e.g. roots developed on stems

aecidiospore – asexually produced, binucleate spore found in the Rust fungi (Order Uredinales)

aecidium – structure in which aecidiospores are produced from the tips of the hyphae; characteristic of the Rust fungi (Order Uredinales)

akinete – single-celled, non-motile, resting spore found in the Blue-green algae (Division Cyanophyta)

alternation of generations – life cycle in which a diploid sporophyte generation, producing haploid asexual spores by meiosis, alternates with a haploid gametophyte generation producing gametes. After fusion of the haploid gametes, the diploid sporophyte is restored

amoeboid – resembling the structure and movement of *Amoeba*, a Protozoan. This type of movement involves the flowing of the protoplasm in a certain direction

amylopectin – a polysaccharide consisting of branched chains of glucose units, making up about 75% of most types of starch

anastomoses – in fungi, cross-connections formed by the fusion of hyphae; characteristic of the Class Basidiomycetes

anisogamous – describes gametes which are unequal in size, but otherwise identical

anisogamy – sexual fusion of unequal gametes

annulus – in fungi, a ring of tissue around the stipe of the fruiting body of some Basidiomycetes

antheridium – a male sex organ or gametangium, producing male gametes or antherozoids

antherozoid – a motile male gamete characteristic of the algae and fungi

aplanogamete – a non-motile gamete

aplanospore – a single-celled, non-motile, asexual spore

apothecium – in fungi, an open, cup-shaped fruiting body, with the hymenium on the upper, exposed surface; typical of some groups of Ascomycetes

archegonium – female sex organ or gametangium, containing one or more oospheres

ascocarp – in fungi, fruiting body characteristic of the Ascomycetes, producing asci

ascogenous hyphae – short, dikaryotic hyphae which are produced from an ascogonium, and on which the asci are formed

ascogonium – in Ascomycetes, a uninucleate or multinucleate female reproductive structure, often with a trichogyne, into which nuclei from antheridia or spermatia enter. The nuclei pair, one male with one female, prior to ascus formation

ascospore – in Ascomycetes, a haploid, elliptical spore formed following fusion of nuclei in an ascus mother cell and subsequent meiosis

ascus (plural: asci) – in Ascomycetes, characteristic structure derived from an ascogenous hypha and producing sexual spores following meiosis; usually found as part of a layer called the hymenium in an ascocarp

aseptate – in algae and fungi, filaments or hyphae lacking dividing walls or partitions

asexual – a form of reproduction in which new individuals are derived from a single parent, without production of fusion of gametes; often involving the formation of buds, spores or gemmae

autospore – an aplanospore having the form of the mature cell in miniature; characteristic of the Order Chlorococcales, Division Chlorophyta

autotrophic – describes organisms which synthesize their own organic requirements from simple inorganic compounds, using light or chemical reactions as sources of energy

auxospore – in Diatoms (Bacillariophyta), a spore often formed after sexual fusion which increases in size before secreting a new cell wall, or frustule, thus restoring the maximum size for the species

basidiospore – in Division Eumycophyta, Class Basidiomycetes, a unicellular, haploid spore formed after the fusion of nuclei in a basidium and subsequent meiosis

Glossary

basidium – in Class Basidiomycetes, characteristic club-shaped structure producing basidiospores; produced on dikaryotic mycelium and often in a hymenium on a fruiting structure

biloproteins – refers to phycobilin pigments; phycoerythrin (red) and phycocyanin (blue); similar molecules to chlorophylls, bound to water-soluble proteins

blepharoplast – the basal body of a flagellum

budding – a type of asexual reproduction characteristic of the yeasts, a group of unicellular Ascomycetes; the new individual is produced as an out-growth from the parent

carotene – especially beta-carotene, an orange pigment found in all photosynthetic plants

carotenoids – a group of yellow, orange, red or brown pigments that absorb light in the blue-violet range

carpogonium – in Rhodophyta (Red algae), the female sex organ produced on the haploid gametophyte generation

carpospore – in Rhodophyta (Red algae), a non-motile, unicellular spore formed after sexual fusion and subsequent division of the zygote

centroplasm – in Cyanophyta (Blue-green algae), the central, clear area of the protoplasm, which contains the chromatin

chitin – a nitrogen-containing polysaccharide forming the main component of the cell walls of fungal hyphae

chlamydospore – in algae or fungi, a thick-walled resting spore

chlorophylls – a group of green photosynthetic pigments common to all photosynthesizing plants; absorbing light in the red and blue-violet ranges

chloroplast – a plastid containing photosynthetic pigments, especially chlorophylls, so that it looks green

chromatin – a complex of DNA and proteins, which makes up the chromosomes; stains with basic dyes

chromatophore – a plastid, or chromoplast, containing photosynthetic pigments

chromoplasm – in Cyanophyta (Blue-green algae), the peripheral region of the protoplasm, which contains the photosynthetic pigments

chromoplast – a plastid containing mostly carotenoid pigments, whose red or yellow colour masks any chlorophylls that may be present

cilium (plural: cilia) – a thread-like organelle, identical in structure to a flagellum but shorter, which beats in a regular rhythm; usually arranged in groups and often associated with locomotion

clamp connection – in Basidiomycetes, a structure which is formed during the division of a dikaryotic cell to ensure that sister nuclei are separated in the daughter cells; a bridge-like hyphal connection characteristic of the secondary mycelium of many Basidiomycetes

cleistothecium – in Ascomycetes, a type of ascocarp which is completely closed; ascospores are released on the decay of the cleistothecial wall

coenobium – in Chlorophyta, especially Order Volvocales, a colony of biflagellate, motile cells connected together by strands of protoplasm, giving some degree of co-ordination. All the cells are embedded in a gelatinous matrix, and the number and arrangement of cells in the coenobium is fixed and definite for each species

coenocytic – having multinucleate cytoplasm surrounded by a cell wall

colony – in algae, a group of unicellular organisms living together, but all independent of one another

columella – in fungi, especially Order Mucorales, a structure which projects into the sporangium, formed by a bulging septum cutting off the sporangium from the sporangiophore

conceptacle – in Phaeophyta, particularly Order Fucales, a flask-shaped cavity in the tip of the lamina containing the sex organs

Glossary

conidiophore – in Eumycophyta, a special hypha from which conidiospores are cut off

conidiospore – in Eumycophyta, an asexual spore consisting of one or more cells, cut off from the tip of a conidiophore

conjugation – the fusion of gametes, usually isogametes, in sexual reproduction

contractile vacuole – found in some unicellular algae, an osmoregulatory organelle which periodically discharges its contents, consisting mostly of water, to the outside of the organism

cortex – in non-vascular plants (i.e. those plants lacking phloem and xylem), the tissues found below the epidermis. In vascular plants, the tissue between the epidermis and the vascular tissue

cyanophycin – in Cyanophyta, a storage protein found in granules in the chromoplasm

cyst – a thick-walled resting stage, which may be formed when conditions are unfavourable for growth

cytoplasm – the living contents of a cell, consisting of all the organelles and inclusions except the nucleus

dendroid – branching in a tree-like manner

dichotomy – division or branching of an organ or structure into two equal portions

dikaryotic – having two nuclei per cell

dioecious – bearing male and female sex organs on separate plants

diploid – any nucleus, cell or organism which has twice the haploid number of chromosomes; often shown as 2n. All zygotes are diploid, as are the sporophytes in any alternation of generations

dorsiventral – having distinct upper (dorsal) and lower (ventral) sides

endoplasmic reticulum – a system of membranes found in the cytoplasm of eukaryotic cells

endospore – in the Cyanophyta, a reproductive spore

epilithic – growing on rock surfaces

epipelic – growing on mud or sand

epiphyte – a plant which grows attached to another plant, but is not parasitic on it

eukaryotic – descriptive of cells in which the nucleus is enclosed in a definite membrane

facultative parasite – an organism which can live as a parasite or as a saprophyte depending on the conditions

filamentous – cells arranged end to end; a chain of cells

fission – asexual division of a unicellular organism into two similar daughter cells; division of the nucleus followed by cleavage of the cytoplasm

flagellum (plural: flagella) – a long, fine, hair-like organelle, whose beating brings about locomotion of an organism or gamete; similar in structure to a cilium but usually only 1, 2 or 4 present per organism or gamete

frustule – in Bacillariophyta, the silica cell wall composed of two valves, the epitheca and the hypotheca, which fit closely together

fucoxanthin – a brown, photosynthetic, xanthophyll pigment, present in the Bacillariophyta and the Phaeophyta

fusiform – spindle-shaped

gametangium – a gamete-producing structure or sex organ

gamete – a haploid sex cell which fuses with another haploid gamete to form a zygote

gametophyte – the haploid generation, which produces gametes

glycogen – a storage polysaccharide, characteristic of animals, but present in some algae and fungi

gonidium – in Volvox, an asexual reproductive cell which divides to give rise to a daughter coenobium

haematochrome – an orange-red, iron-containing pigment

haploid – any nucleus, cell or organism having a single set of chromosomes; often shown as n

haustorium – in parasitic fungi, an absorptive hypha or outgrowth which penetrates the cells of the host to take up nutrients

Glossary

helical – twisted into a coil

heterocyst – in Cyanophyta, a large, thick-walled, empty cell found in many filamentous genera; filaments break up at heterocysts, thus providing a means of vegetative reproduction for the species

heterosporous – having spores of two sizes, microspores and megaspores

heterothallic – condition in which fusion of gametes from the same individual, or strain, cannot occur; cross-fertilization is necessary, between two compatible thalli

heteromorphic – having more than one growth form; in an alternation of generations, the sporophyte will differ morphologically from the gametophyte

heterotrichous – having two types of growth in an individual plant; an erect, or upright, system and a prostrate system of filaments

holdfast – in algae, the structure which anchors the plant to the substratum

homosporous – having spores of one size only

homothallic – condition in which gametes from the same individual or strain are able to fuse; the thallus is self-fertile

hormogonium – in Cyanophyta, a short portion of a filament which may become detached and grow into new plant

hormospore – in Cyanophyta, a multicellular, spore-like body with a very thick wall

hyaline – clear, transparent

hymenium – in Eumycophyta, the fertile layer in a fruiting body, consisting of asci in the Ascomycetes and basidia in the Basidiomycetes, together with sterile hairs, or paraphyses

hypha – in Eumycophyta, one of the filamentous structures making up the plant body or thallus

hypnospore – in algae, an asexual spore with a very thick cell wall

hypocotyl – that part of the stem of a seedling between the root and the cotyledons

intercalary – inserted between others

internode – the portion of a stem, or axis, between two nodes

isodiametric – of cells, having equal dimensions all round; of equal height, width and length

isogamous – describes gametes which are equal in size; morphologically identical

isogamy – sexual fusion of equal gametes

isomorphic – having similar form or shape; in an alternation of generations, both generations are morphologically identical

lamina – in algae, the flattened portion of the thallus

laminarin – main storage carbohydrate found in the Phaeophyta (Brown algae); composed of linked glucose units

littoral zone – region of the sea-shore between high and low tide levels

mannitol – an alcohol which forms a major carbohydrate reserve in the Phaeophyta

medulla – the central part of an organ or tissue; the central mass of hyphae in the stipe of members of the Order Agaricales, Division Eumycophyta

meristem – a group of undifferentiated cells retaining the ability to divide, resulting in growth and the formation of new tissues

monoecious – with both male and female sex organs borne on the same plant

monospore – in algae, an asexual spore produced singly; germinates to give rise to a new plant

mucilage – a slimy, complex carbohydrate substance produced by many algae

mutant – the result of a mutation or change in the structure of the genes or chromosomes, producing a change in the characteristics of an organism

mycelium – in fungi, the mass of hyphae forming the plant body or thallus

mycobiont – in lichens, the fungal component

mycorrhiza – a symbiotic association between a fungus and the roots of a vascular plant

neutral spore – type of asexual spore found in some groups of algae

node – position on a stem or axis at which leaves or branches arise

nucleolus – a small body lying within the nucleus, rich in ribonucleic acid (RNA)

nucleus – the organelle, present in eukaryotic cells, which contains the chromosomes

obligate parasite – a parasite which cannot live independently of its host

oidium (plural: oidia) – in fungi, an asexual spore produced by the breaking up of a hypha into individual cells

oogamy – the sexual fusion of two dissimilar gametes; the male gamete being small and usually motile with very little stored food, and the female gamete being large and non-motile with larger food reserves

oogonium – in algae and fungi, a female sex organ containing one or more oospheres, or female gametes; unicellular

oosphere – a naked, spherical, haploid female gamete; an ovum or egg cell

oospore – a fertilized oosphere which develops a thick wall before dormancy or a resting phase

organelle – a part of a cell having a definite structure, bounded by a membrane and having certain specific functions: e.g. nucleus, chloroplast

osmoregulation – the process by which some unicellular algae which possess contractile vacuoles maintain the correct concentration of water and ions in their cells

ostiole – the opening, or aperture, of a conceptacle typical of the Order Fucales in the Phaeophyta, or of a fruiting body typical of the Class Ascomycetes in the Eumycophyta

palmate – having finger-like lobes

palmella – in algae, a phase occurring in some motile, unicellular genera when water is scarce; the cells mass together, withdraw their flagella and become embedded in a gelatinous matrix

pantonematic – applied to the 'flimmer' or 'tinsel' type of flagellum, which has tiny projections or hairs along its length

papilla – a small projection or protuberance

paramylum – an insoluble carbohydrate related to starch; food reserve typical of Euglenophyta

paraphysis – sterile hair consisting of a single row of cells; associated with the oogonia in the Order Fucales of the Phaeophyta

parasite – an organism that lives on or in another living organism, called the host, from which it derives its food. The host does not benefit in any way and may be harmed by the parasite

parenchymatous – composed of parenchyma tissue; consisting of unspecialized cells with thin cellulose cell walls and living contents

parietal – attached to, around or near the walls of the cell

parthenogenetically – reproduction in which a female gamete can develop into a new individual without being fertilized by a male gamete

pectic – made up of polysaccharide substances which commonly form the middle lamella between plant cell walls

pellicle – in Euglenophyta, the external, elastic, proteinaceous layer surrounding the organism; allows flexibility while conferring definite shape

perennial – a plant that continues to grow from year to year

peridium – in fungi, the outer wall of the spore-producing structure; may be simple as in the Myxomycophyta, or made up of interwoven hyphae as in some Eumycophyta, especially in the Class Ascomycetes

peripheral – around the outside; surrounding

periphysis – an unbranched, sterile hair situated near, or lining, the ostiole in the perithecia of some Ascomycetes; or developed in the upper portion of conceptacles in Order Fucales of the Phaeophyta

periplast – in Euglenophyta, the exterior portion of the cytoplasm; may be flexible, allowing the cell to change shape, or rigid

Glossary

perithecium – in Class Ascomycetes, a flask-shaped ascocarp with a hymenium lining the inner surface. When mature, ascospores are ejected through the ostiole at the apex

phagotrophic – feeding by the ingestion of extracellular particles

phialide – in Eumycophyta, Class Ascomycetes, a special branch at the tip of a conidiophore from which conidiospores are produced in succession; an alternative term for a sterigma

phialopore – in Chlorophyta, Order Volvocales, a pore at one pole of a developing coenobium

phloem – that part of the vascular tissue of higher plants, composed of living cells, through which organic substances are transported

photoreceptor – a sense organ or organelle responsive to light

phototactic – movement shown by an organism in response to variation in light intensity or to the direction of light

phototrophic – describes organisms which use light energy to synthesize their organic requirements from simple, inorganic compounds; photosynthetic

phycobiont – in lichens, the algal component

phycocyanin – a blue photosynthetic pigment, a biloprotein; gives the characteristic colour to the Cyanophyta (Blue-green algae)

phycoerythrin – a red photosynthetic pigment, a biloprotein; gives the characteristic colour to the Rhodophyta (Red algae)

pileus – in Eumycophyta, the upper part, or cap, of the fruiting body or basidiocarp; especially descriptive of the Basidiomycetes

plakea – in Chlorophyta, Order Volvocales, a descriptive term for the curved plate of 8 cells which develops during the formation of a daughter coenobium

plankton – collective term for minute plants (phytoplankton) and animals (zooplankton) which occur at the surface of fresh and salt water

plasmodium – a multinucleate, amoeboid protoplasmic mass which forms the vegetative phase of members of the Myxomycophyta

plurilocular – in Phaeophyta, a zoosporangium divided up into compartments in which single zoospores develop

podetium – in lichens, a stalk-like structure on the thallus which bears the apothecia

polarity – differentiation in structure or function between two ends of an axis

prokaryotic (or procaryotic) – describes an organism or cell whose genetic material is not confined within a true nucleus bounded by a distinct membrane

prosenchyma – a tissue composed of loosely-packed interwoven filaments or hyphae; forms the thallus in some Phaeophyta and some Eumycophyta

prostrate – growing flat on the substratum. In some Chlorophyta, that portion of the thallus from which erect branches grow

protoplast – the living contents of a cell excluding the cell wall

pseudoparenchymatous – describes a tissue consisting of filaments or hyphae, often closely interwoven, which lose their identity, their cells becoming isodiametric so that the tissue in its mature state closely resembles parenchyma

pycnidium – in Eumycophyta, Class Ascomycetes, an open cup- or flask-shaped cavity lined with spores

pycnospore – an asexual spore formed within a pycnidium. It resembles a conidiospore

pyrenoid – in algae, a small protein structure found in the chloroplasts, associated with the formation and storage of starch

rhizoid – a uni- or multi-cellular filamentous outgrowth which anchors a plant lacking roots to its substratum

rhizomorph – in Eumycophyta, Class Basidiomycetes, a stout strand composed of compacted hyphae lying parallel

saprophyte – an organism which obtains its food from dead or decaying organic matter

saptotrophic – feeding like a saprophyte; applied to colourless members of the Euglenophyta which may take in organic molecules

Glossary

septate – having cross walls or septa

septum – a dividing wall or partition in a plant structure

sexual – a form of reproduction in which two haploid cells (gametes) or two haploid nuclei fuse to produce a diploid cell, a zygote

sieve cell – a cell, typically found in phloem, which has a protoplast but lacks a nucleus

silicified – impregnated with the element silica; especially found in cells of the Bacillariophyta

siphonaceous – descriptive of some algae, particularly in the Order Siphonales, where the thallus is a coenocyte

somatic – descriptive of non-reproductive cells and cell divisions

soredium – in lichens, a vegetative spore which consists of a few algal cells and fungal hyphae

spermatangium – in Rhodophyta, a male gametangium producing spermatia

spermatium – in algae, a non-motile male gamete, characteristic of the Rhodophyta. In fungi, a non-motile, uninucleate male structure which may fuse with a female structure; found in certain groups of Ascomycetes and Basidiomycetes

spermatozoid – a motile male gamete; an antherozoid

spermogonium – a flask-shaped structure in which spermatia are formed; especially characteristic of the Class Basidiomycetes, Order Uredinales

sporangiophore – an erect hypha or stalk which bears a sporangium

sporangiospore – a non-motile, asexual spore produced in a sporangium

sporangium – an asexual reproductive structure in which sporangiospores develop

spore – a reproductive structure, usually small, consisting of one or more cells; may be asexual or sexual

sporophore – in Basidiomycetes, a spore-bearing structure or fruiting body

sporophyte – the diploid generation in a life cycle, producing haploid spores after meiosis

sporulation – the act of producing spores

stellate – star-shaped

sterigma (plural: sterigmata) – in Basidiomycetes, a small projection which develops from a basidium and on which a basidiospore is produced; in Ascomycetes, an alternative term used for a phialide, found in some genera of the Ascomycetes

stigma – an eyespot; a light-sensitive organelle found in some flagellate algae, and in some motile male gametes

stipe – a stalk; in Basidiomycetes, part of the fruiting body; in Division Phaeophyta, the portion between the holdfast and the lamina

stolon – in Mucorales, a hypha which grows horizontally across the substratum, enabling rapid spread of the vegetative thallus

substratum – the material or substance on which an organism grows

suspensor – in Mucorales, the side branch of a hypha which supports a gametangium during conjugation

symbiosis – an association between two organisms which benefits both partners

taxonomy – the study of the classification of plants and animals according to their similarities and differences

teleutosorus – in Heterobasidiomycete fungi, an aggregation of dikaryotic hyphae producing teleutospores

teleutospore – in Heterobasidiomycete fungi, a dikaryotic spore from which basisia and basidiospores develop after fusion of nuclei

thallus – a simple plant body showing no differentiation into stem, leaf and root; typical of the algae, fungi and lichens

trichogyne – in Ascomycete fungi, the long, hair-like neck of an ascogonium which is receptive to the spermatia or conidiospores

trichome – in Cyanophyta, a single row of cells which, when surrounded by mucilage, is termed a filament

Glossary

unilocular – in Phaeophyta, describes a type of sporangium in which the contents undergo meiosis to produce haploid zoospores

uredosorus – in Heterobasidiomycete fungi, an aggregation of dikaryotic hyphae producing uredospores

uredospore – in Heterobasidiomycete fungi, a binucleate asexual spore

utricle – in Chlorophyta, particularly Order Siphonales, a swollen lateral branch arising from the densely interwoven filaments which make up the erect thallus. The utricles contain chromatophores and are photosynthetic

vascular tissue – in higher plants, the conducting tissues, composed of xylem and phloem

vesicle – in Phycomycete fungi, a bubble-like structure into which zoospores are released, or in which zoospores are developed

volutin – a polymer of inorganic phosphate which may occur as granules in the cytoplasm of micro-organisms

xanthophylls – a group of carotenoid pigments which absorb light in the blue-violet range; found in all photosynthetic plants

xylem – a composite tissue in higher plants which contains water-conducting tissue and which contributes to the mechanical support of the plant

zoosporangium – a structure in which zoospores are formed

zoospore – a motile asexual spore

zygospore – a zygote which develops a thick, resistant wall, enabling it to survive a period of dormancy

zygote – a diploid cell which results from the fusion of two gametes